MW01055793

Praise for
Dam It! Electrifying America and Taming Her Waterways

"*Dam It!* is mesmerizing. The book's many photos make the story jump off the pages. You don't have to be a veteran of the business to be fascinated by the intriguing details of the history and impact of hydro plants. Happy reading!"

Warren Witt, Director of Hydro Operations, Ameren

"Underwood has written a lovely primer on the origins of the electricity industry and a very good survey of the development of its hydroelectric portion. I particularly appreciate his fairness to one of the great figures of the industry, Samuel Insull, who is all too often treated as only a robber baron."

John Rowe, Former CEO of Commonwealth Edison and Exelon

"No technical degree required to enjoy this lively, and often quite personal, account of the building of an industry and of the nation. Full of fascinating insights, fast-paced, and well-written."

Barry Posner, Chair of Department of Management
and Entrepreneurship, Santa Clara University

"Powered by his gift for storytelling, Underwood weaves a riveting narrative about America's quest to harness the power of water through hydroelectricity. It is a tale of historic significance that reads like a novel—full of spellbinding characters, political intrigue, and inventive genius. I love this book."

Mike Leonard, Emmy Award-winning television journalist

"*Dam It!* highlights the major contributions of dam designers and builders in bringing light and energy into the modern age. It makes us appreciate what remarkable structures dams are and the immense challenges pioneering engineers, entrepreneurs, and policy makers overcame in their quest to harness falling water and power America."

Ernest Freeberg, Author of
Age of Edison: Electric Light and the Invention of Modern America

Praise for
Dam It! Electrifying America and Taming Her Waterways

"This book is really good. It tells an important, fascinating story and is spellbinding. The Afterword profiling Underwood's dam builder grandfather made me verklempt."

–Ann Crane, Pulitzer Prize-winning journalist

"Underwood personalizes the great projects and people who electrified America through managing water power. He captures the intense, raw, down-and-dirty competition between General Electric and Westinghouse, brings to life the significant movers of the industry, and gives a behind-the-curtain look at the powerful organizations and cutthroat politics that shaped the playing field. Well written and entertaining."

–William J Martin, CEO, CME Energy

DAM IT!

DAM IT!

Electrifying America and Taming Her Waterways

ROBERT L. UNDERWOOD

COLOMA
PRESS
CHICAGO • BOSTON

Dam It! Electrifying America and Taming Her Waterways
Copyright © 2023 by Robert L. Underwood

COLOMA
PRESS
CHICAGO • BOSTON
www.colomapress.com

INTERNATIONAL STANDARD BOOK NUMBER (ISBN):
979-8-9852452-0-2, HARDCOVER. 979-8-9852452-1-9, ELECTRONIC BOOK

The author is available for speaking events. Contact Coloma Press for information about author appearances or for information about quantity discounts for *Dam It!*

Book and cover design by Adams Press, Evanston, Illinois

 This book is manufactured from recyclable materials.

PRINTED IN THE UNITED STATES OF AMERICA

10 9 8 7 6 5 4 3 2 1

PUBLISHER'S CATALOGING-IN-PUBLICATION DATA
(Prepared by Five Rainbows Cataloging Services)

Names: Underwood, Robert L., author.
Title: Dam it! : electrifying America and taming her waterways / Robert L. Underwood.
Description: Winnetka, IL : Coloma Press, 2023. | Includes index.
Identifiers: LCCN 2022911358 (print) | ISBN 979-8-9852452-0-2 (hardcover)
Subjects: LCSH: Water-power. | Dams. | Waterways. | Engineering. | BISAC: TECHNOLOGY & ENGINEERING / Civil / Dams & Reservoirs.
Classification: LCC TK1081 .U86 2023 (print) | LCC TK1081 (ebook) | DDC 621.31/2134--dc23.

In loving memory of my grandfather, George P. Jessup:
"A dam engineer and proud of it."

Table of Contents

Introduction

We live in an electrified world—a world where electricity is a readily available commodity that we take for granted. Electricity powers our lives. We can plug in anywhere, anytime and know we will receive the power needed for our lights, our hair dryer, our phone, our television, and other daily necessities.

This wasn't always so. Scientists experimented with electricity as early as the 1700s, but it was Thomas Edison's demonstration of the first commercially viable electric light bulb in 1879 that liberated society from near-total dependence on daylight and spawned the use of electricity in almost every aspect of our lives today. In fact, the electric light usually is ranked among the innovations that most have changed history, right up there with the wheel, the printing press, and the steam engine.

Edison's discovery launched the race to electrify America. Electrification drove the American economy from 1900–30. Electric utilities and their suppliers utilized more capital than any other industry. By the early 1930s, seventy percent of homes in the United States had electricity.

Tapping the water in our rivers has been an integral part of generating the electric power that America needs. The mechanical power of falling water is an age-old tool. Dams had been used extensively to tame rivers for flood control, to create reservoirs to store drinking and irrigation water, and to generate mechanical power via water wheels. It was only natural that dammed water be used to generate electricity. And used it was as the electric power industry exploded.

Both the mechanical power of water and the other major energy source for electricity generation—steam produced by burning coal—converted their energy into electrical current by driving spinning turbines. Where rivers and streams could be tapped, waterpower was the cheaper energy source. By 1940, more than fifteen hundred hydroelectric facilities produced about one third of America's electrical energy.

The story of the evolution of hydroelectric power rivals that of any transformational technology we have seen arise from Silicon Valley. Eccentric inventors, financial wheeling and dealing, political intrigue, mind-boggling engineering and construction feats, inspiring personal stories ... it's all here.

Chapter One

In the Beginning

Thomas Edison's 1879 invention of the commercially viable incandescent electric light bulb revolutionized the world. It marked the end of the "dark ages". Before then, most human activity occurred during daylight hours. At night, all interiors were either dark or dimly lit. Illumination required a device with a flame, whether it be candle, whale oil lamp, kerosene lamp (introduced in the 1850s), or gaslight. None of these sources of illumination burned brighter than about one twelfth the intensity of today's 100-watt light bulb. All flickered, created smoke and odor, and were a constant fire or explosion hazard. Because they consumed oxygen, they also could cause breathing problems in poorly ventilated rooms.

At the time of Edison's invention, America was starkly different from the country we know today. Four factors were at work radically reshaping the nation: settlement of the vast geographic expanse from the Mississippi River to the Pacific Ocean, the migration of large numbers of people to America, the wave of industrialization and related urbanization, and a mammoth investment in a national grid of railroads. Furthermore, the nation still was reeling from the impact of the Civil War. It was only in 1877—twelve years after the official end of fighting—that the last US Army troops were removed from the South. The region remained poverty-stricken and was nearly totally dependent on agriculture for its economy.

The entire United States was far less populated than it is now. The population, however, was growing rapidly. In 1880, the US population was 50 million people, about 15 percent of the current population. It would rise another 13 million people in both of the next two decades. Over 5 million immigrants arrived in the 1880s decade alone. They often found jobs in mills and factories and were an enabling factor in the country's urbanization and industrial transformation.

Almost 75 percent of Americans lived in New England, the middle Atlantic, and the north–central states. The country still was largely rural. Its urban population was concentrated primarily in the New York City-to-Philadelphia corridor. Only twenty cities had more than one hundred thousand people.

Westward expansion of the country was in full swing. The hammering of the Golden Spike in 1869 completed the country's first transcontinental railway, instantly ended the isolation of California and the Great West from the eastern half of the United States, and opened vast areas to settlement and economic development. Settlers from the East, lured by the promise of cheap or even free land and by the natural resources of the West, poured across the Mississippi River to mine, farm, and ranch. They needed reliable supplies of water for drinking and irrigation and to power mining operations.

Settlement from the East transformed the Great Plains. The Homestead Act, passed by Congress in 1862, had opened the Great Plains to settlers, giving 160 acres of land to any person who would live there for five years. Farmers plowed the natural grasses to plant wheat and other crops. These crops were shipped via rail to eastern markets. The cattle industry also rose vastly in importance as the railroad provided a practical means for getting cattle to market.

The western movement and the simultaneous annihilation of the huge herds of bison that had roamed the plains threatened the lives and livelihoods of the Native Americans living in the West, and they resisted. In 1876, Lt. Colonel George Custer and his cavalry were massacred by combined forces of the Lakota, Northern Cheyenne, and Arapaho tribes in the Battle of the Little Bighorn. The resulting public outcry accelerated efforts to control Native Americans and remove them from their ancestral lands. By the 1880s, most tribes had been confined to reservations.

Meanwhile, the discovery of deposits of natural resources in various parts of the West hastened its development and led to land rushes, boom towns, and extensive mining operations. Gold was discovered in California in 1848. Stimulated by the ensuing Gold Rush, San Francisco became an urban island, the nation's ninth largest city by 1880. Silver was discovered in Nevada in 1859, and copper in Arizona and Montana in the 1870s and early 1880s. Coalfields were developed in Colorado, New Mexico, and Wyoming.

As the railroad grid in the West expanded and transported more and more raw materials, products, and people, communities sprang up along the rail routes and the urbanization of the West began. The urban population for the entire Western region would exceed 50 percent of its total population by 1920.

Although the spread of railroads across America in the 1870s was transforming the way people and goods traveled around the country, the principal modes of transportation remained horses, waterways, and human foot power. Prior to the development of the nation's rail network, waterways had been the most efficient mode of transportation where they could be accessed. The opening of the Erie Canal in 1825 had slashed the time to move people and bulk goods from New York City 500 miles to the Great Lakes. The Ohio, Mississippi, and Missouri rivers also had become important transportation arteries. Railroads provided faster intercity passenger and bulk freight movement than waterways. Transportation within cities and between small towns, however, remained highly dependent upon horses.

One of the world's fastest-growing cities, Chicago became the nation's transportation hub during the 1870s, linking East and West. The main rail lines from the East ended in Chicago, and those oriented toward the West began there, so the city became the nation's principal trans-shipment and warehousing point. It was a processing center for natural resource commodities extracted in the West. Hogs and cattle were shipped to slaughterhouses there, with the meat then shipped to eastern cities. After publication of the first Montgomery Ward mail-order catalog in 1872, Chicago also became home to national retailers that used the rail lines radiating from Chicago to deliver goods to rural America. No longer isolated or dependent on the local general store for the limited goods it offered, urban and rural families alike suddenly had a vast range of goods available to them at standardized, relatively low prices.

Trees and water had fueled America's growth since colonists first arrived in the 1600s. The East was heavily forested, and wood was used extensively as a building material and as an energy source for heating. As colonists pushed westward and settled farther inland along rivers and streams, they put waterwheels to work in sawmills and gristmills. These mills were critical to the fledgling nation's economy. Over time, dams and waterpowered mills proliferated in number and complexity and spawned the country's industrial transformation during the 1800s. By 1840, the twenty-six states that comprised the United States had about 65,000 dams—an amazing one dam for every 261 people.[1]

In the 1800s, with the East becoming deforested, coal became an attractive alternative to wood as an energy source for heating and power generation. It provided more heat per pound than wood and was much more portable. Industrialization and steam-powered machinery stimulated rapidly growing demand. Expansion of the nation's railway network after the Civil War vastly increased possibilities for moving it from mines to far-flung usage points. By the late 1880s, the amount of energy produced nationally from coal exceeded the amount produced from wood.

Industrialized manufacturing had begun in the early 1800s in New England, where wealthy merchants built textile mills (and mill towns to support them) along the region's rivers. Falling water provided the mechanical power necessary to drive machines. Specialized, repetitive manufacturing tasks assigned to wage laborers replaced handicraft production by artisans at home, irrevocably changing the nature of work. Factories quickly sprang up along rivers throughout the Northeast and upper central states. Textile weavings, shoe making, leather tanning, papermaking,

---

[1] Martin Doyle, *The Source: How Rivers Made America and America Remade Its Rivers,* New York, W.W. Norton & Company, 2018, p. 220.

hat making, clock making, and gun making all were mechanized. This industrialization led to product standardization, greatly reduced production costs, and the ability to match output more effectively to changes in product demand.

Workers were needed to manufacture all these industrialized goods. Young girls, enticed by the prospect of relatively high, steady, cash wages, left farms to work in mills. By the time of the Civil War, nearly a thousand textile factories in New England alone employed more than one hundred thousand people. But working conditions in factories of this era were dismal. Most factory workers toiled twelve to fourteen hours a day, five and a half days a week, performing monotonous tasks. In winter, when the sun set early, oil lamps lit factory floors. Employees strained to see their work and coughed as rooms filled with smoke from the lamps. Mills often were unbearably frigid in the winter and hot and humid in the summer. Fires occurred frequently. Workers' hands and fingers were maimed or severed when caught in machines. Limbs or entire bodies were crushed.

Industrialization accelerated after the Civil War. The wave of industrialization and the country's mushrooming rail network required never-before-seen capital investments. Wall Street and its proverbial financial tycoons emerged to meet the challenge. Astute entrepreneurs and their financial advisers realized that consolidating given industries not only could provide economies of scale, but also could eliminate competitors and facilitate price- and wage-setting to maximize profitability. Unprecedented mergers followed, largely unfettered by government intervention.

Reality hit hard with the Financial Panic of 1873 and the wrenching, seemingly endless depression that followed until about the time of Edison's electric light bulb invention. The Panic was triggered by the bankruptcy of Jay Cooke & Company, the country's premier banking house. After essentially financing the Civil War for the Union, Cooke had turned to financing railroads through bonds that promised fixed returns to investors. Rampant investment speculation, coupled with rising interest rates, made it impossible to make required bond payments. The Panic's run on banks and the subsequent decimation of stock and bond values were brutal. The New York Stock Exchange closed for ten days. During the ensuing depression, hundreds of banks closed. Construction of new rail lines, formerly one of the backbones of the economy, plummeted. Some eighteen thousand businesses failed. Average wages fell by nearly 25 percent. A collapse in food prices caused great poverty in rural America. Many former Civil War soldiers became transients. Relief rolls exploded in major cities, with 25 percent unemployment in New York City alone.

Unemployed workers demonstrated in Boston, Chicago, and New York in Winter 1873–74, demanding public work. In New York's Tompkins Square, police charged the crowd of demonstrators and pummeled thousands of men and women with clubs. The most violent strikes in American history followed the Panic, including a strike in Pennsylvania's coalfields in 1875, led by the secret labor group known as the Molly Maguires. Masked workers exchanged gunfire there with Pinkerton's Coal and Iron Police, a private force aligned with coal field owners and commissioned by the state. During a nationwide railroad strike in 1877, mobs destroyed railway hubs in Pittsburgh, Chicago, and Cumberland, Maryland.[2]

---

[2] Scott Reynolds Nelson, "The Real Great Depression," *The Chronicle Review*, October 17, 2008, p. B98.

During these trying and even desperate times, workers came to distrust both business owners and government, seeing them as a common, colluding adversary. The federal government during Ulysses S. Grant's two terms as president, from 1869–77, was unapologetically pro-business and notoriously corrupt.

Wealthy industrialists such as Andrew Carnegie (steel), Cornelius Vanderbilt (shipping and railroads), and John Rockefeller (oil), aided by their financiers (especially J.P. Morgan), solidified their hold over their industries and increased their governmental influence as a result of the Panic and following depression. This was the Gilded Age, a time when materialistic excesses collided with extreme poverty. The industrializing American economy was generating unprecedented wealth for a small minority of people while many factory workers and farmers struggled just to survive. Three distinct social classes arose in urbanized America—the very wealthy, a middle class of factory and business owners and managers, and low-income laborers. The disparity in wealth distribution has been exceeded only in recent years.

The industrial and financial aristocracy lived in palatial homes and indulged in opulent amusements. The Vanderbilts built a 154-room mansion in New York City. Meanwhile, working class families subsisted in squalid and overcrowded nearby tenements. The Gilded Age of haves and have nots led directly to worker unions, curbs on monopolies and other competitive abuses, and government regulation of business practices. These changes would contribute to the evolution of hydroelectricity in years ahead.

Dwelling units circa 1880 essentially were islands, largely unconnected to the outside world. They universally lacked indoor plumbing, running water, waste disposal, and central heating. All water—whether pumped from a well or a cistern or occasionally obtained from an urban street hydrant for cooking, dishwashing, bathing, laundry, and housecleaning—had to be carried in and the dirty water carried back out after use.

The first indoor plunger-type toilet was invented in 1875. Most families relied on chamber pots or outdoor privies for decades, however. Most heat came from fireplaces using wood or coal. Bedrooms seldom were heated. People bathed infrequently since taking a bath required carrying a tin or wooden tub into a room where it could be filled with water that had been heated on a kitchen stove or in a fireplace.

The housewife's life was exceptionally difficult. She bore the daily task of moving clean and dirty water, coal, wood, and ashes in and out of the house, as well as preparing and cleaning up after daily meals. She typically hand-made the clothes the family wore, often using unsized paper patterns. She repeatedly heated a flat iron on the kitchen stove or in a fireplace to iron clothes. Little did she know how electricity would change her life.

All in all, this was an era of transition in America that pointed to exciting changes and accelerating economic development in the years ahead. Americans were viewed worldwide as being inventive, having a talent in mechanical engineering, for devising machines to ease the toil of workers and farmers, and for honing manufacturing methods. They did big things. In the last three months of 1879, the incandescent electric light, the first reliable internal combustion engine, and wireless transmission all were invented. Within the prior three years, the telephone and phonograph had been invented. America would be transformed in the decades ahead, and hydroelectric power would play a leading role in making that happen. It indeed is a dam good story.

Fig. 1.1. Driving the Golden Spike Completing
the First Transcontinental Railroad
May 10, 1869

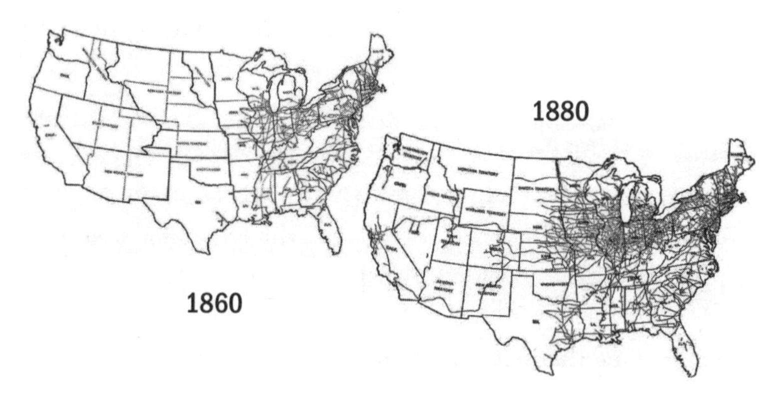

Fig. 1.2. US Railroad Lines in 1860 and 1880

Fig. 1.3. Textile Mill

Fig. 1.4. Homestead Act Settler Family in Nebraska

Fig. 1.5. Vanderbilt Mansion in New York City c. 1880

Fig. 1.6. New York City Tenements

Chapter Two

The Dawn of Electric Power

Thomas Edison invented the practical incandescent light bulb. And he launched what became today's electric power industry, forever tying us to a central source of energy. Electricity now is ubiquitous. It powers our lives. As we shall see, his world-changing invention also stimulated an amazing era of dam building in order to generate hydroelectric power in the ensuing rush to electrify America .

It all started in 1878 when Edison devoted himself and his Menlo Park, New Jersey, laboratory to developing a commercially viable electric lighting system. His extraordinary ability to conceptualize entire systems, invent critical missing building blocks, and then develop and optimize the overall system enabled him to build and install a complete, viable electric lighting system in New York City's financial district in 1882. Not only did he invent the long-lasting incandescent light bulb in 1879, but he also made vast improvements in electric generator (dynamo) technology needed to effectively power those light bulbs, realized that networking light bulbs via parallel rather than series circuits was crucial to having successful lighting systems, and designed and installed the central power station and wiring grid necessary to support the incandescent bulbs placed in service.

Edison did not start from scratch in his drive to develop a commercially viable electric lighting system. The first electromagnetic generator, the Faraday disk, was built in 1831 by British scientist Michael Faraday. Generator technology advanced thereafter. In 1841, Frederick de Moleyns of England was granted the first patent for an incandescent lamp. His design used powdered carbon heated between two platinum wires contained within a vacuum bulb. It was far too expensive, the inside of the bulb blackened easily, and the lamp had too short a filament life to be practical. Numerous light-bulb experiments subsequently were conducted by others in search of an inexpensive, long-lasting filament. The economic, long-lived electric bulb was the missing link that would spark the indoor lighting revolution and create market demand for significant electric power systems.

Thirty-one-year-old Edison was the ideal person to tackle this challenge and make electric lighting happen. He accumulated 1,039 US patents during his lifetime, more than anyone else ever. Among his previous inventions were the quadruplex telegraph in 1874, which made it possible to simultaneously transmit and receive four telegraph messages over a single wire, and the phonograph in 1877. His Menlo Park laboratory became the first industrial R&D laboratory. It was perfect for multifaceted, systems research and development.

Edison understood how to achieve technology breakthroughs and convert them into commercial successes. Once he set his sights on a goal, he was completely and relentlessly driven and focused on that goal. He was famous and, because of his successes, was able to attract substantial amounts of capital. He also was a savvy promoter and an often-cutthroat competitor.

In 1878, as Edison began his electric lighting quest, he started by securing the funding necessary for his intended efforts from the Vanderbilts and J.P. Morgan. He quickly decided that any electric lighting system had to be economically competitive with the gas lighting systems of the day. That led him to conclude that it was important to minimize the amount of copper required for the conductors in generators and transmission lines in electric power systems. The laws of physics showed that that could be achieved by using high-resistance bulb filaments and operating those bulbs at relatively low voltage (approximately 100 volts). He further determined that parallel circuitry was necessary in the power distribution network so that a power interruption at one place would not cause more interruptions throughout the network. (Parallel circuitry is what keeps all your holiday string lights glowing even when one bulb goes out; series circuitry ruins the whole string if one bulb goes out.)

After numerous experiments with a variety of filaments, Edison in October 1879 had finally created a successful test of a bulb: It featured a coiled carbonized carbon thread filament that lasted nearly fifteen hours. He filed for a patent the next month, which was granted amazingly quickly on January 27, 1880. This was the first commercially practical incandescent light bulb. Several months later, he discovered an even-better carbonized bamboo filament that could last more than twelve hundred hours.

Edison made the first public demonstration of his incandescent light bulb on December 31, 1879, during a Menlo Park open house. Lights had been placed around the laboratory property and inside the laboratory buildings. The Pennsylvania Railroad ran special trains to Menlo Park on the day of the demonstration in response to the great excitement the event created. More than three thousand people came by train, carriage, farm wagon, and horseback and were awestruck by the brilliant display.

In 1879, the team at Menlo Park also made tremendous progress in the design and construction of a generator with low internal resistance tailored to work efficiently with the high-resistance bulbs Edison was pursuing and to minimize the amount of required copper.

Soon it was time to target a showcase first installation of the complete electric power system. In this, Edison confidant and advisor Grosvenor Lowrey played an important role. A New York attorney who, as general counsel for Western Union, had worked closely with Edison on telegraph patent litigation, Lowrey had close contacts within the New York financial and political worlds. It was he who had secured Edison's funding from the Vanderbilts and Morgan.

Edison, with Lowrey's input, set his sights on installing a complete electric power system in lower Manhattan. He envisioned a central electric power station from which electricity would be distributed as a utility to the public over a one-square-mile area in New York's financial district. There was a high density of banks, brokerages, and offices there that could afford to replace existing gas lights with decidedly superior electric lighting. In addition, the area housed offices of existing investors (e.g., Drexel, Morgan and Company) and of many other potential investors. The offices of *The New York Times* also were there. The central power station would burn coal to turn water into steam to drive reciprocating steam engines to provide the motive power to electric generators. The generated electricity would be distributed via a grid of underground transmission lines servicing the entire square-mile area.

With the general initial installation project plan in place, Edison undertook an extremely detailed canvass of the targeted service area. When it was completed, he knew how many gas jets there were in every building, the average number of hours they burned daily, the owner of each jet, the location of every elevator to which a motor might be applied, and much more. This enabled him to lay out and size the underground transmission grid. In August 1880, he applied for a patent on the unique "feeder and main" system he devised.

He soon discovered, though, that New York City politicians were not enamored with his proposal to dig up the streets in lower Manhattan to install the needed nearly 100,000 feet of wiring. His proposal for a system franchise was denied. Behind the opposition of some aldermen undoubtedly were gaslight interests and fears of lamplighters who might lose their jobs.

Lowrey swung into action and unleashed a lobbying tour de force. In December 1880, he arranged to have a special train bring the mayor and aldermen to Menlo Park. They arrived at dusk to see bulbs bathing the exteriors and interiors of the laboratory buildings with light. After a tour and demonstration, someone on cue said it was time to go upstairs for a drink. The group was led to the darkened second floor. A switch was flipped, and lamps suddenly went on all around the room to reveal an utterly lavish spread from New York's famous Delmonico's restaurant. Not only were the setting and dinner spread impressive, so was the fact that the entire room could be illuminated instantaneously by flipping a remotely located switch. This was not possible with any other kind of interior lighting system then available. To emphasize that point, the room lights were switched off and then a moment later back on again a second time. After cocktails, dinner, and plenty of wine, Lowrey introduced Edison to much fanfare and then presented the case for the Edison franchise. This did the trick, and the franchise was forthcoming in April 1881.

Work to install the lower Manhattan power system began at a furious pace. Edison purchased buildings at 255–257 Pearl Street in the shadow of the Brooklyn Bridge to convert into the central station. The necessary steam boilers, steam engines, generators, and auxiliary equipment to be housed there were built or procured, installed, and tested.

Invention continued in finding ways to adequately insulate conductors to be deployed in the underground network, to meter electricity usage, and for myriad other system details. Installation of the underground network began in late Fall 1881, was suspended when the ground froze during the winter, and resumed in late February 1882. In Spring and Summer 1882, six constant voltage "Jumbo" dynamos were installed at Pearl Street, each capable of powering 1,200 incandescent lamps. They were driven by reciprocating steam engines supplied by four coal-fired boilers. On September 4 of that year, the power system was put into commercial operation. This was the first permanent central-power-station-based public electric utility in the world.

The day after commercial operation commenced, *The New York Times*, an initial customer itself, stated in an article that:

> The light was soft, mellow, and grateful to the eye and it seemed almost like writing by daylight to have a light without a particle of flicker and with scarcely any heat to make the head ache ... The decision was unanimously in favor of the Edison light lamp as against gas ... The Edison electric light has proved in every way satisfactory.

On the first day of operation, there were fewer than 90 customers using a total of around 400 incandescent lamps. Within a year, 513 customers using about 10,000 lamps were being served.[1] The system was Edison's high-profile showpiece to entice others to purchase both franchises and equipment.

Just twenty-six days after coal-fed Pearl Street went live, far away from New York City in Appleton, Wisconsin, the first Edison waterpowered electric plant (i.e., hydroelectric plant) in the world went into operation on September 30, 1882.[2] Appleton at first blush would seem an unlikely pioneering Edison installation. Nonetheless, it did make sense. A town with about ten thousand residents located along the Fox River 30 miles southwest of Green Bay, Appleton was in the center of the booming papermaking industry.

A process for making paper from wood pulp rather than from rags was introduced to Wisconsin in 1872. Wisconsin's abundant forests, ample water supply, and ability to transport logs via lakes and riverways made it ideal for paper production. The industry exploded. When installation of electricity was being considered in 1882, Appleton already had a community gas lighting system. It also had paper mills with waterpowered beater[3] rooms. There was an existing dam across the Fox River and a canal system originally built to provide hydromechanical power in the mills. Another key ingredient was the town's prosperous

---

1 See, e.g., http://ethw.org/Milestones:Pearl_Street_Station,_1882.

2 This was not the first hydroelectric plant. There had been a handful of other plants powering arc lighting and/or incandescent lamps that were developed independent of Edison in England by Joseph Swan. See *Hydro Review*, October 1997, pp. 46–47.

3 Beaters are blending machines that process wood fibers into a water-based pulp and mix additives into the stock.

and progressive citizenry. Furthermore, as would be expected, Edison's business representatives were eager to install an electric system here, if possible, to expose another part of the country to the virtues of incandescent lamps.

The Western Edison Electric Light Company was incorporated May 25, 1882, for licensing of Edison systems in Illinois, Wisconsin, and Iowa. The company was the predecessor of Chicago's Commonwealth Edison Company, about which we will hear much more.

H.J. Rogers, a prominent and entrepreneurial Appleton citizen who was president of both the Appleton Paper and Pulp Company and the Appleton Gas Light Company, became keenly interested in the electric light while on a fishing trip with his friend, H.E. Jacobs, who had become a representative for Western Edison upon its formation. It turned out that Jacobs was quite a salesman. Although he had no ability then to demonstrate the new lighting system, he arranged for an Edison engineer to come to Appleton in July 1882 to explain the system to Rogers and three other Appleton businessmen whom Rogers had assembled.

The group decided on the spot to test the possibilities of electricity for lighting their mills and their homes and for providing electric lighting to other Appleton businesses and homeowners. Rogers clearly saw that if electric lights were going to replace gas lighting, controlling electric lighting in Appleton could offset any negative impact on his Appleton Gas Light Company. Another motivator for Rogers undoubtedly was the fact that he was building for his wife Cremora a magnificent mansion on a bluff overlooking the Fox River and the mills below. Installing electric lamps in the house would fit with his desire to feature the best and newest in the mansion.

Rogers and his associates formed the Appleton Edison Light Company and were granted the exclusive rights to Edison technology in the entire Fox River Valley. On August 18, they ordered two Edison "K" type dynamos. Each would be capable of lighting about two hundred fifty incandescent bulbs and would be driven by water power.

Things progressed quickly. For an initial test, one of the dynamos was placed in the beater room of Rogers's Appleton Paper and Pulp Company building, where it was connected directly via gears and belts to a water wheel that was driving the mill's beaters. Lamps were placed in that plant, in another paper mill about a mile away, and throughout Rogers's new mansion. Pole-mounted electric lines directly connected the two remote locations to the dynamo. On September 30, twenty-six days after Edison's Pearl Street system had commenced operations, the initial Appleton test system became operational. The local newspaper declared that the illuminated buildings were "as bright as day."

Problems surfaced almost immediately. Due to varying loads carried by the paper-mill beaters, the speed of the water wheel and dynamo varied greatly. Sometimes the voltage became so high that all the lamps being powered by the dynamo burned out; at other times the bulbs dimmed dramatically. Several weeks later, this was alleviated by moving the dynamo to a lean-to attached to the building with its own separate water wheel. There were no voltage regulators or fuses. Short circuits abounded because the distribution lines were bare copper and because wiring within the structures was lightly insulated with a cotton covering. All these problems were dealt with. It is remarkable that all of this was being accomplished by a franchisee far away from Edison and the insights and solutions he undoubtedly could have provided.

By November, the second dynamo that had been ordered was installed at another site, the Vulcan Hydroelectric Central Station. This plant incorporated a unique and sophisticated water supply system to regulate the water pressure driving the plant's water wheel. By year's end, the complete system was lighting five mills, a blast furnace, and three residences. In January 1883, Appleton's Waverly House became the first hotel in the Midwest to be electrically lighted in every room. Thereafter, service expanded quickly. By 1886, Appleton even had an electrified street railway.

The Appleton system, albeit initially quite primitive, demonstrated the practicality of using water power for generation of electricity, indicated why it would be highly desirable in many situations, and paved the way for the critically important role that hydropower was to play in electrifying America. An obvious advantage was that, after system installation, operating costs could be dramatically lower than for coal-fed systems such as Edison's Pearl Street station in New York. Under full headway, the boilers in the Pearl Street station consumed about five tons of coal daily, for an annual cost of about $125,000 in today's dollars.[4] Material handling added significant additional cost. On the other hand, the water that powered the Appleton system was free, renewable, and clean.

The Appleton system also took advantage of extensive existing infrastructure that had been developed for providing hydromechanical power. This made installation relatively simple and inexpensive.

Early electrical power systems were direct current systems and, as a result, were highly localized. Direct current transmis-

---

4 Thomas P. Hughes, *Networks of Power: Electrification in Western Society, 1880-1930*, Baltimore, MD: Johns Hopkins University Press, 1983, pp. 39 and 43.

sion line losses restricted operations in practice to a distance of about a mile from the generating system. Hydroelectric system developers thus took advantage of river falls within the centers of communities and existing hydromechanical power apparatus whenever possible. Areas with mills and factories and a high number of residences owned by prosperous people were targeted especially. Electric lighting for homes was expensive, but quickly became an exciting, safe, and coveted luxury. Monthly charges for electric service were competitive with monthly charges for gas light service, but an incandescent lamp bulb cost $1.00–$1.50[5] (about $22–$35 in today's dollars) and all too often burned out due to shorts and power surges. Factories also began to have electric lighting and to use electric motors to drive machinery.

Prior to 1885, direct current motors usually had less than one horsepower capacity and, thus, were limited in application. The first reliable and efficient direct-current motors with greater than one horsepower capacity were developed by Frank Sprague, a former Edison employee. Edison encouraged use of these motors in factories because daytime use there would complement relatively high nighttime illumination loads in people's homes. And, as in Appleton, he similarly encouraged their use for electric trolley systems in urban areas. These electrically driven street cars could be controlled more easily and operated more cheaply than their horse-drawn counterparts and greatly reduced the horse-waste problem where they operated: The average horse produced twenty to fifty pounds of manure and a gallon of urine daily.

Direct-current hydroelectric system installations in America in the 1880s evolved into a patchwork quilt of dedicated spe-

---

5 See, e.g., http://ethw.org/Pearl_Street_Station.

cial-purpose systems and small-footprint community utilities located in sections of the country where abundant water power could be harnessed. Because local conditions varied and since technology was changing quickly as the issues that system promoters and operators encountered in practice were addressed and overcome, there was limited standardization among these pioneering installations.

The number of installations grew rapidly. *Electrical World* reported that, by August 1886, between forty and fifty electric light plants "largely or wholly dependent on waterpower" were on line or under construction in the United States and Canada. Two hundred electric companies in the United States used waterpower to generate some or all of their current by 1889.[6] In addition, there were countless other special-purpose installations where generators were connected directly to factory water wheels.

The race was on. The world had glimpsed the life-changing potential of electricity via incandescent lighting and limited other initial uses. That was exciting enough, but dreamers, schemers, and tinkerers already were devising many other possibilities. Sure, there were obstacles. So what? This was going to be big.

---

[6] Duncan Hay, *Hydroelectric Development in the United States, 1880-1940*, Washington, DC: Edison Electric Institute, 1991, pp. 15-16.

Fig. 2.1. Edison at Pearl Street Next to 100 kW Generator

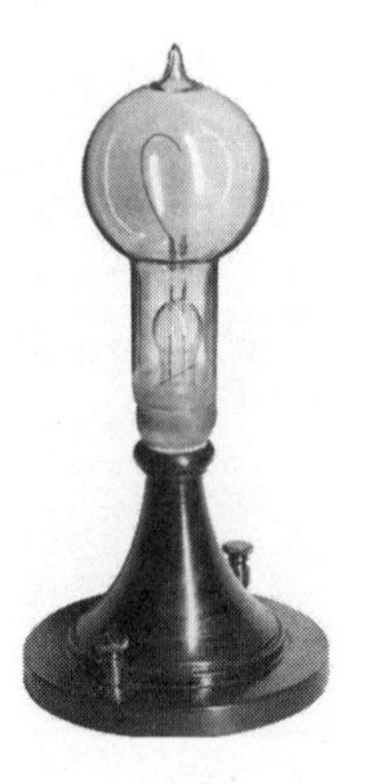

Fig. 2.2. Original Edison Light Bulb from 1879

Fig. 2.3. Edison Confidant Grosvenor Lowrey

Fig. 2.4. Pearl Street Station

Fig. 2.5. Appleton's H.J. Rogers

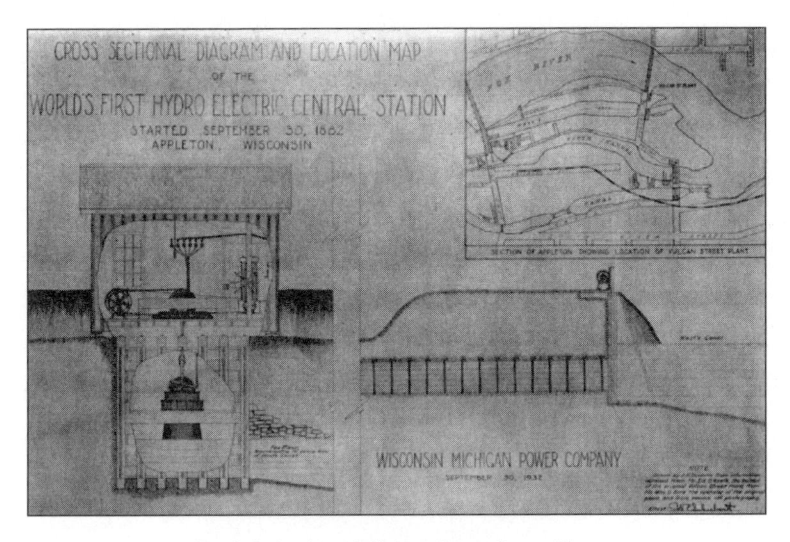

Fig. 2.6. Appleton's Vulcan Street Power Plant

Chapter Three

Power Struggle

Edison's electrical power system was revolutionary. Nonetheless, as we have seen, it had one critical drawback. Because it was a direct-current system, transmission line losses limited its practical operation to about one mile from the system's electricity generators. There had to be a way to eliminate this drawback, and one soon emerged.

Direct current is a continuing flow of electrons in a single direction. It is the type of current produced when the conducting lead wires from a load such as a light bulb are connected to the two terminals of a battery as shown in Figure 3.1. This is the type of current Edison had worked with since his early days in telegraphy. It was natural that he would build his initial electrical power systems around the technology.

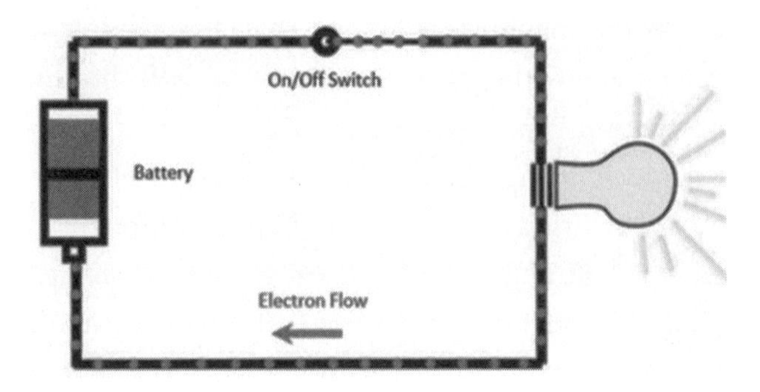

Figure 3.1. Direct Current Circuit

Direct-current systems, however, work at constant voltage throughout the system. Electricity leaving the generator at 110 volts, the typical standard for circuits in the United States today, remains at 110 volts save for line losses due to heat and friction as current passes through transmission lines to electric light bulbs, motors, or any other loads on the system. The smaller and longer the wire, the more electrons bump into each other and into the walls of the wire. This creates friction and heat, diminishing the wire's current capacity. If the amount of current a wire carries exceeds its current capacity, the wire can overload and melt and cause a fire.

According to basic laws of physics known to Edison and his associates but still not universally understood or accepted at the time, the same amount of power will be transmitted through a conducting wire if the voltage is raised from 110 volts by some factor and the current is reduced by the same factor. The reduced current allows a smaller and lighter transmission wire to be used. The laws of physics additionally state that the energy loss in a conducting wire is proportional to the square of the current traveling in it times the length of the wire divided by the wire's cross-sectional area. It follows that, for example, doubling the voltage allows the same size wire to transmit the same amount of power four times the distance. Recognizing these facts was key to unlocking the distance limitation of Edison systems.

The solution to the range limitation of direct-current systems emerged through the development of alternating-current systems. Although direct current (DC) had evolved from roots in battery power, alternating current (AC) had its roots in the esoteric physics of electromagnetism. Alternating current describes the flow of charge that periodically changes direction. It had its genesis in British scientist Michael Faraday's 1831 experiments probing possible interactions between magnetism and electrical current. He observed that when a changing magnetic field is applied to a conductor, an electromotive force is induced. This causes current. He constructed a crude electromagnetic generator that year.

In simplest terms, generators produce alternating current as a loop of wire is spun inside of a magnetic field, thereby inducing a current along the wire. Because the wire spins and periodically enters a different magnetic polarity, the voltage and current alternate on the wire. A slip ring with spring-loaded brushes taps the power off the rotor and keeps it flowing consistently.[1] A simple AC generator is depicted in Figure 3.2.

---

1 DC generators and AC generators create current basically the same way. DC generators use rotating switches called commutators to disconnect power during the reverse current portion of the wire loop's cycle. From Faraday's time until Edison's light-bulb invention, generator technology advanced with a primary emphasis on direct current. Direct current was easier to understand and most applications in those days were for direct current.

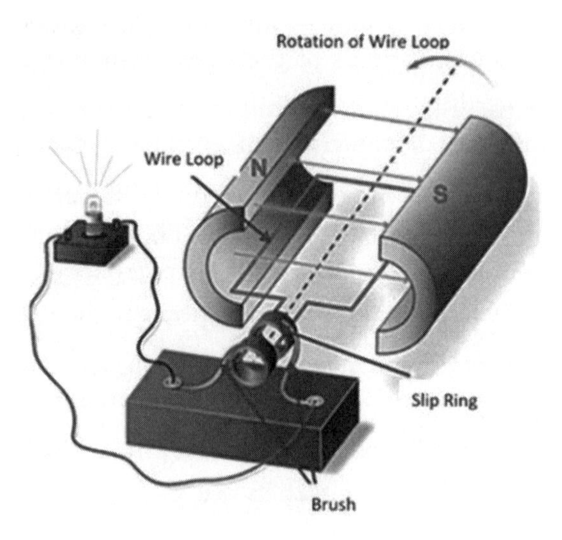

Figure 3.2. Simple AC Generator

Faraday's other major alternating-current advance was the transformer. When he discovered electromagnetic induction, he did so by manipulating an "induction ring." He wound each side of a thick iron ring with a separate insulated wire and found that varying the electric current in the coil on one side of the ring (the primary coil) would induce a current in the coil on the other side of the ring (the secondary coil). This was the first electrical transformer. Soon thereafter, it was determined that the ratio of the primary voltage to the secondary voltage varied as the ratio of the number of primary wire winding turns to the number of secondary winding turns. Transformers thus could step up or step down the voltage in an alternating-current circuit.

In September 1882, nine days after Edison's Pearl Street lighting system became operational in New York City, Frenchman Lucien Gaulard and his English business partner John Dixon Gibbs patented a complete system that combined alternating current with transformers of their own design as a means to solve the long-distance power transmission issue. Over the next several years, they demonstrated their system in various installations, including a 50-mile transmission circuit in Turin, Italy, in 1884. Their groundbreaking work quickly was improved upon by others, most notably in transformers developed and introduced commercially in 1885 by engineers at the Ganz Works in Budapest, Hungary. The Ganz system was widely adopted. By 1890, about seventy central stations of various sizes were in operation in Europe using the company's AC generators, transformers, and controls.

It was George Westinghouse who brought commercial use of alternating current to America, in the mid-1880s. Four months older than Edison, he, too, was a serial inventor, a successful owner of multiple manufacturing companies, and a consummate entrepreneur. Not part of the New York establishment, he was, rather, a prominent citizen of Pittsburgh, the industrial powerhouse known for its iron and steel mills, railroad equipment factories, and access to transportation westward on the Ohio River and eastward via the Pennsylvania Railroad.

At the age of nineteen, he obtained the first of the 361 patents he eventually would receive. Four years later, he revolutionized the railroad industry by inventing the rail car air brake and founding the Westinghouse Air Brake Company, thereby becoming a wealthy man. Later, as rail traffic increased and railroad yards emerged, Westinghouse saw the need for better signaling devices and interlocking switches. Using purchased patents and his own inventions, he developed a complete electrical and compressed-air signal system and, in 1881, founded another major business, Union Switch and Signal Company.

Three years later, he drilled a natural gas well on the expansive property surrounding his mansion and invented an entire delivery system, for which he secured thirty-six patents. He then began a natural-gas utility to supply thousands of homes in the Pittsburgh area. The system relied upon his invention of a reduction valve that allowed high-pressure gas from a well to be delivered at low pressure at the point of use.

All three of these Westinghouse enterprises involved transmission over distance—compressed air for air brakes, electricity for railroad signals, and gas for his natural-gas utility. Each also involved some crucial linking mechanism to connect the transmitted substance with the rest of the system.

In 1884, Westinghouse became intrigued with electricity and foresaw the potential for explosive growth in electric lighting. By chance, while he was surveying the field and simultaneously dealing with an electrical signal lamp question at Union Switch and Signal, his brother introduced him to William Stanley, Jr.

Stanley, a young inventor and electrician described as a "live wire," had worked for two other electric lighting companies and owned his own small business. Westinghouse employed him to work at setting up a light bulb facility and developing a DC system. Westinghouse also paid Swan Incandescent Electric Light Company to retrieve DC dynamo and incandescent lamp patents that Stanley previously had assigned to it. The new Westinghouse DC system debuted later in 1884.

In Spring 1885, Westinghouse had a sudden inspiration upon reading in a British technical journal a description of an AC system employing Gaulard-Gibbs step-down transformers. He realized that this could be the solution to the distance limitations of Edison-type DC systems. He promptly bought an option on the Gaulard-Gibbs American patent rights and ordered a transformer as well as a commercially available Siemens AC generator typically used to power arc-lighting systems.

The transformer arrived in Pittsburgh in terrible condition late in November, accompanied by Reginald Belfield, an English Gaulard-Gibbs employee. Westinghouse invited Belfield to be a house guest at his estate while he rebuilt the transformer. The two conversed nightly about the transformer and future developments. By the end of December, Westinghouse had completely redesigned and improved the transformer, making it manufacturable and efficient. In January 1886, he formed the Westinghouse Electric Company to pursue the immense AC electric systems opportunity he envisioned and exercised his option to purchase the American Gaulard-Gibbs patent rights.

Meanwhile, Stanley had fallen ill, ostensibly from Pittsburgh's famously terrible smog, and moved to Great Barrington, Massachusetts, for health reasons. Westinghouse renegotiated their arrangement and redirected Stanley's efforts to setting up a test AC electric system in Great Barrington. Belfield joined Stanley there early in January, whereupon they constructed six of the redesigned transformers, strung 4,000 feet of copper transmission lines through the town center, installed the Siemens generator, and readied a steam engine to power it. The transformers were hidden in wooden boxes and installed out-of-sight in the basements of buildings that would be lighted.

The system was tested successfully in mid-March, and several dozen customers were gained by month end. By mid-April, Stanley had designed a more reliable generator to replace the original Siemens generator. Westinghouse then had another complete system installed and thoroughly tested in Pittsburgh before the inaugural commercial installation in November 1886 of a 498-lamp system in Buffalo's leading department store.

Westinghouse, unlike Edison, did all of this with no publicity or promotion. Nonetheless, he swiftly secured orders for systems in numerous locations.

Until this point, the only real application for high-voltage transmission had been arc-lighting systems. Arc lamps were extremely bright and used for lighting streets, factory yards, and large interior spaces. They were not suitable for normal interior lighting. Arc lamps needed high voltage (typically above 3,000 volts), and some ran better on AC. Arc lighting was maintenance intensive; it buzzed and flickered; and the high voltage made it dangerous without proper precautions. Because arc-lighting transmission lines in urban areas often were strung on poles shared with telegraph, telephone, and other electrical wires, they posed a real danger if they were mishandled, dangled, or broken.

Early AC systems such Westinghouse's initial installations, although a breakthrough, still had deficiencies. They lacked a practical motor. In addition, AC meters did not exist, AC systems were more difficult to control than DC systems, and high-voltage operation did create safety issues.

By the end of 1886, upon hearing about Westinghouse's Great Barrington trial system installation and learning the details of Westinghouse's first commercial AC system installation in Buffalo, Edison perceived a serious threat. Westinghouse was a formidable rival with a new, potentially disruptive technology, access to large amounts of capital, and a record of impressive achievements. Edison's patents could not block Westinghouse from pursuing his revolutionary work. What Edison could not understand, though, was how Westinghouse could be successful given the dangers of high-voltage transmission. 110-volt DC was safe. It could give a person a jolt, but it was not lethal. High-voltage AC (or DC!) could kill people.

The business threat to Edison became critical during 1887. By the end of the year, Westinghouse had sixty-eight AC central stations constructed or under contract. At that time, despite having been in business considerably longer, Edison had only 121 DC central stations installed or under contract. Edison's problem was exacerbated when a European syndicate cornered the world copper market and drove up prices. DC central plant installations required around three times as much copper as AC station installations. Something had to be done.

By coincidence, Edison received correspondence late in 1887 from a member of the New York State Death Commission seeking his opinion as a pre-eminent authority on electricity about using electrocution for executing convicted criminals and asking for information about "the necessary strength of current to produce death with certainty in all cases and under all circumstances."[2] After several gruesome botched hangings that led to widespread public outcry, the Governor of New York had appointed a commission to find a form of execution more humane than hanging. In December 1887, Edison replied:

> The best appliance in this connection is, to my mind, the one which will perform its work in the shortest space of time and inflict the least amount of suffering upon its victim. This, I believe, can be accomplished by the use of electricity, and the most suitable apparatus for the purpose is that class of dynamoelectric machinery which employs intermittent currents. The most effective of these are known as "alternating machines", manufactured principally in this country by Geo. Westinghouse ... The passage of the current from these machines through the

2 Terry Reynolds and Theodore Bernstein, "Edison and 'The Chair,'" *IEEE Technology & Society Magazine*, March 1989, p. 20.

> human body, even by the slightest contacts, produces instantaneous death.[3]

Soon thereafter, in February 1888, Edison seized upon his patent position and his reputation, the safety concern, and arguments that AC systems were inferior to DC systems to publicly go on the attack. In the forceful, red-covered, eighty-three-page booklet, *A WARNING FROM THE EDISON ELECTRIC LIGHT CO.*, Edison assaulted AC systems, his lighting system competitors, and Westinghouse in particular. The publication emphatically stated that, "It is a matter of fact that any system employing high [voltage], i.e., 500 to 2,000 [volts] jeopardizes life." It went on to detail the horrific deaths of numerous workers by high-voltage AC. The vitriol continued by saying, "All electricians who believe in the future of electricity ought to unite in a war of extermination against cheapness in applied electricity, wherever they see that it involves inefficiency and danger." The booklet pointedly asserted that George Westinghouse was "the inventor of the vaunted system of distribution which is today recognized by every thoroughly-read electrician as only an *ignis fatuus* (i.e., a will-o'-the-wisp or deceptive hope), in following which the Pittsburg company have at every step sunk deeper in the quagmire of disappointment."[4]

AC system element refinements and technology advancements meanwhile were occurring quickly. In April 1888, a Westinghouse engineer developed an induction meter that allowed the company to determine how much AC electricity a customer

---

[3] *Ibid.*, p. 21.

[4] Quotations from *A WARNING FROM THE EDISON ELECTRIC LIGHT CO.*, a pamphlet published by the company in February 1888, pp. 26, 70, and 72.

used. In 1887 and 1888, a number of inventors and scientists in Europe and America focused independently on the AC motor challenge. In a case of what scholars call the phenomenon of simultaneous discovery, the challenge was met, most notably by Italian Galileo Ferraris and Serbian-American Nikola Tesla.

Ferraris had conceived of a rotating magnetic field, lectured about it in March 1888, and built an operating motor. Tesla filed a group of fundamental patent applications late in 1887 for not only a brushless polyphase AC motor employing a rotating magnetic field but also a complete power system adapted to such a motor. On May 15, 1888, two weeks after the first seven of Tesla's patents were granted, he demonstrated a complete polyphase system during a lecture to a convocation of the American Institute of Electrical Engineers.

AC systems up until this point had been single-phase systems. A polyphase motor uses several out-of-phase, or polyphase, currents to establish a rotating magnetic field in its stator windings. Polyphase motors are self-starting. By 1900, polyphase systems would eclipse single-phase systems, and the three-phase system would become the *de facto* standard.

Westinghouse again struck decisively in 1888, first purchasing Ferraris's American patent rights and then, in July, entering into a generous agreement with Tesla to license his patents and to provide him funds to further develop his AC induction motor and polyphase system. Westinghouse now had all the elements of a complete AC system to compete head to head with Edison's DC system.

Tesla reported that when he was working for Edison in 1884 and had tried to interest him in pursuing his AC induction motor concept, Edison told him "very bluntly that he was not interested in alternating current; there was no future to it and anyone who dabbled in that field was wasting his time; and besides, it was a deadly current whereas direct current was safe."[5]

In mid-1888, when solutions to the AC system usage metering and motor deficiencies had been demonstrated, DC proponents began to focus intensely on the lethality of alternating current. In June, Harold P. Brown, a previously obscure, self-professed electrical engineer, published a letter to the editor in New York's *The Evening Post* calling alternating current "fatal" and "damnable" and asserting that it submitted the public "to constant danger from sudden death." He contended that, even at low voltages, alternating current was much more lethal than direct current, and he proposed legislation outlawing all AC above 300 volts to prevent "the wholesale risk of human life."[6] This would have legislated commercial AC electrical systems out of existence.

Attacked by engineers associated with AC systems for having no data to support his charges that alternating current was more lethal than direct current, Brown quickly appealed to Edison for support for a series of experiments to gather evidence for his assertions. To this day, historians never have been able to prove that there previously was a connection between the two men. *The Evening Post*, though, was owned by a major original Edison investor, Henry Villard, who soon would become president of Edison's electric company. Edison immediately offered Brown use of his laboratory and the assistance of two key employees to conduct his experiments.

By the end of July, Brown had dispatched stray puppies and grown dogs captured in the vicinity of the New Jersey Edison

---

[5] John J. O'Neill, *Prodigal Genius: The Life of Nikola Tesla*, New York: Cosimo, Inc., 2006, p. 62.

[6] Harold P. Brown, Letter to the editor, New York: *The Evening Post*, June 5, 1888, p.7.

laboratory and was ready to show the world his evidence of AC's lethality. In a highly publicized demonstration in a lecture room at Columbia College, attended by seventy-five members of the electrical community and the press, he proceeded, to the intensifying alarm of the audience, to subject a 76-pound dog to increasing levels of DC. At 1,000 volts, the dog contorted in excruciating pain. Many horrified spectators left the room. Brown then administered 330 volts AC to the terrified animal, which collapsed dead. Shocked audience members shouted their derision and angrily demanded that he cease such torture. It was too late, of course, for the dog.

Unshaken, Brown returned four days later to dispatch three more dogs with 330-volt AC. One 53-pound dog survived for four awful minutes before finally expiring. But Brown did not stop there. In December, at Edison's laboratory complex, witnessed by reporters, members of the New York State Death Commission, several physicians, and Edison himself, Brown used 700-volt AC to electrocute in turn two calves and a healthy 1,230-pound horse. He used animals of this size to demonstrate conclusively that AC could be used to execute criminals. Soon thereafter, New York State adopted "death by alternating current." Via this bizarre and underhanded campaign, Edison and his cohorts managed to make "AC" and "danger" synonymous in the public eye, diverting attention from DC systems' distance limitations and higher cost.

In May 1889, New York State sentenced an axe murderer named William Kemmler to be its first criminal to die by AC electrocution. The destitute criminal's appeal, asserting that this method of execution was nothing short of cruel and unusual punishment, was handled by a legal all-star, the flamboyant and expensive W. Bourke Cockran. It was widely speculated, but never proven, that Cockran had been engaged by Westinghouse.

Edison appeared in person to testify in opposition to the appeal, saying that AC most certainly could kill instantly and painlessly (the animals dispatched by Brown undoubtedly would have disagreed!). His fame and stature carried the day, and the appeal was denied in October.

After additional unsuccessful appeals, Kemmler's electric-chair death was to be conducted on August 6, 1890. Brown and Edison had conspired to obtain a used Westinghouse AC generator for the event after Westinghouse refused to sell a new one for the purpose. The execution did not go well. Kemmler somehow was still alive after the first attempt. Power was administered for several minutes in a second attempt. Kemmler literally was fried. The next day, *The New York Times* headline screamed, "FAR WORSE THAN HANGING; KEMMLER'S DEATH PROVES AN AWFUL SPECTACLE." Thereafter, Brown was not heard from again.

Edison and his associates used the public perception they had been building to go even further. In November 1889, Edison had stated in an article he published, "My personal desire would be to prohibit entirely the use of alternating currents. They are as unnecessary as they are dangerous."[7]

He began attempts to legally ban AC. The initial attempt was in Virginia. In February 1890, Edison was the first witness to appear before the State Senate. Opposed by local arc-lighting companies, Edison's proposed legislation was soundly defeated when it was characterized as nothing but a dogfight between large "Yankee" companies. Legislative attempts elsewhere in the United States and in Canada also were futile.

---

[7] Thomas Edison, "The Danger of Electric Lighting," *The North American Review*, November 1889, pp. 625–633.

Another weapon that Edison wielded in his campaign against AC was his extensive patent portfolio, particularly his fundamental light-bulb patents. His goal was to make potential purchasers of systems other than his own DC system afraid that in the future they either could be shut down or face large financial exposure. Edison typically licensed his central-station customers and proclaimed the strength of his patents. Licensees expected that the patents would convey monopoly power to them. As early as 1885–86, Edison began suing both unlicensed utility companies and competing lamp manufacturers for patent infringement. His February 1888 *WARNING!* booklet contained a threat aimed squarely at potential Westinghouse system users:

> Suits have been instituted by the Edison Company upon these patents against [sellers] of what is called the Westinghouse system (which so-called system we are informed is the one you contemplate using), in the United States Circuit Court for the District of New Jersey. Should the decisions in these suits be favorable to the Edison Company, they will at once proceed to obtain injunctions throughout the country against all infringers of their patents and will at the same time bring action to recover damages for past infringements.[8]

Numerous patent suits and counter suits filed by the various protagonists in the battle between AC and DC wound their way through courts in America and Europe. Costs of the suits became staggering. Gaining control of specific patents even became a significant factor in the mergers and acquisitions of leading players as the industry increasingly consolidated.

---

8 *A WARNING FROM THE EDISON ELECTRIC LIGHT CO.*, February 1888, p. 41.

Despite Edison's attempts to impede Westinghouse and AC systems, Westinghouse was winning the one battle that mattered most: system installations. In fact, September 1890, just one month after the gruesome Kemmler electrocution, was a banner sales month for Westinghouse. Then October's many orders included a 6,000-light system for Baltimore and 1,500-light systems for both Lincoln, Nebraska, and Elmira, New York. By the end of that year, in just four years since his initial AC system installation, Westinghouse had installed a total of three hundred central stations with a capacity of half a million incandescent lamps.

Edison's employees and affiliates increasingly urged him to embrace AC and add AC systems to his systems offerings. During the August 1889 annual meeting of Edison utility companies, managers from many cities complained that they were being crushed by competition from Westinghouse and passed a resolution asking the parent Edison company to provide "a flexible method of enlarging the territory which can be profitably covered from their stations for domestic lighting by higher [voltages] and consequently less outlay of copper than that involved by the three-wire [DC] system."[9] Edison remained obdurate.

Financial considerations eventually determined the outcome of the AC-versus-DC battle. Electrical-systems manufacturing was highly capital intensive, requiring enormous amounts of money to develop complete product offerings, build major factories, and establish national sales and service networks. By 1889, three companies had emerged as dominant players in the industry—Edison General Electric, Westinghouse Electric, and

[9] George Westinghouse, "A Reply to Mr. Edison," *The North American Review*, December 1889, pp. 663–664.

Thomson-Houston—and they were in brutal competition with each other for sales.

Thomson-Houston had been formed in 1883 by shoe-company investors based in Lynn, Massachusetts, and led by Charles Coffin to further develop an arc-lighting system co-invented by Elihu Thomson and Edwin Houston. By 1889, under Coffin's leadership, the company had evolved via a series of patent licensing and sharing agreements, acquisitions, and internal developments to provide arc lighting, AC and DC incandescent lighting systems, and electric street railway systems. Edison believed Coffin was deceitful and underhanded and said he had "boldly appropriated and infringed every patent we use."[10] Westinghouse also disliked and distrusted Coffin. He saw him as having "a very swelled head" and as an aggressive wheeler-dealer who "will make a man about ten different propositions in ten minutes." He was revulsed when Coffin confided that he intentionally had run the price of the stock of his company down to deprive Thomson and Houston of financial gains from a new stock issue.[11]

In April 1889, Edison's significant resources had been brought together by Henry Villard into a single new entity, the Edison General Electric Company. Villard—a railroad tycoon who built a railroad empire in the Pacific Northwest, an international financier with excellent financial connections in his native Germany, and a friend of J.P. Morgan—had been an original investor in Edison's companies. Even before Edison's original Pearl Street central-station system became operational in 1882, Villard had prevailed upon Edison to install the first isolated incandescent

---

[10] Matthew Josephson, *Edison: A Biography*. New York: McGraw-Hill, 1959, p. 360.

[11] Jill Jonnes, *Empires of Light*. New York: Random House, 2004, pp. 234–235.

lighting plant aboard his steamship, *Columbia,* in May 1880.[12] Villard admired Edison and was an active investor and promoter on his behalf. The two men were visionaries with little patience for day-to-day business matters and financial constraints.

Villard loved big deals. In 1886, he had been chartered by Deutsche Bank to invest a large pool of German money in American enterprises. Not surprisingly, Edison's companies were a prime investment target. Edison's business was expanding rapidly, and so were the capital needs of his companies. More than half of the new capital infused into Edison General Electric Company upon formation was furnished by Villard's German sources. J.P. Morgan and his partners also invested. Villard became president of the streamlined new company but left day-to-day management to Samuel Insull, Edison's highly capable and versatile assistant. Edison was delighted with these events. He received a large cash payout while still retaining an ownership interest in the company and then was able to focus more on his technical endeavors. He wrote gratefully to Villard, "I have been under a desperate strain for money for 22 years and ... one of the greatest inducements was the sum of cash received, so as to free my mind from financial stress, and thus enable me to go ahead in the technical field."[13]

The company continued to focus on incandescent DC lighting systems and contracted with Edison's new, greatly expanded laboratory at West Orange, New Jersey, for further product development. Insull hoped that Edison would develop an AC lighting system and an electric trolley system, but there was no

---

[12] Demand for these isolated plants far exceeded expectations. Ninety-nine isolated Edison plants already were in operation when Pearl Street went online as a central-station-based electric utility.

[13] Josephson, *Edison: A Biography*, p. 354.

guarantee that Edison would put these priorities ahead of his own goals.

Villard did not stop there. Whereas Edison and Westinghouse believed in competition and survival of the fittest, Villard and Coffin personified the era's robber-baron mentality. They believed that the way to wealth was to consolidate businesses within a given market in order to create a monopoly, thereby achieving economies of scale and eliminating price competition. The two frequently discussed a possible merger of Edison General Electric and Thomson-Houston almost from the day Edison General Electric was formed. Their bankers initially were unable to determine satisfactory terms, and Edison himself was dead set against it.

Even so, Villard continually worked on cultivating relationships with Westinghouse and Coffin with the goal of ultimately merging their three companies. He had little success with Westinghouse, but early on started sharing sales, production, and earnings information with Coffin. In 1889, they agreed that one would not bid on a contract in Washington, DC, if the other would not compete for one in Richmond, Virginia. Soon they were price fixing. In 1891, they met to negotiate the bids that each would submit for four separate upcoming street-railway contracts.

Meanwhile, Westinghouse faced a financial crisis. He had invested a significant amount of his personal wealth in his company, but he also had borrowed heavily to fund expansion. Most of his financing came from local Pittsburgh sources. It turned out to be a recipe for disaster. The near collapse in November 1890 of London's venerable Baring Brothers & Company banking house triggered a severe recession and financial panic, causing creditors to call their loans to Westinghouse. In his scramble to find new money to avert bankruptcy, Westinghouse had to turn to bankers in New York and Boston. Bankers behind Edison General Electric and Thomson-Houston tried maneuvers to either wrest control of the company from Westinghouse as a condition for providing capital or to have his refinancing fail in order to force bankruptcy. Either of these maneuvers would have cleared a roadblock to an eventual merger of Westinghouse Electric with its rivals.

To help save the company, Tesla relinquished his lucrative rights to future royalties from his AC inventions. He said to Westinghouse, "You have been my friend, you believed in me when others had no faith ... You will save your company so that you can develop my inventions."[14] Eventually, in what many considered a miracle, Westinghouse achieved his necessary refinancing in July 1891 and saved his company—and himself—from ruin.

Not long thereafter, J.P. Morgan stepped in and began negotiations to merge Thomson-Houston and Edison General Electric. He was impressed that, despite the companies having almost equal annual sales, Thomson-Houston was twice as profitable. The negotiated terms left Coffin and his managers in total control of the merged companies. Coffin was to be president of the new company, which was to be named General Electric. The only Edison person slotted for a significant post was Samuel Insull. He was offered the position of second vice president. He instead resigned to become the president of fledgling Commonwealth Edison in Chicago.

Morgan forced Villard to resign from Edison General Electric before the pending merger was announced in February 1892. Edison himself first learned of the impending merger only hours

[14] Jonnes, *Empires of Light*, p. 229.

before it was announced publicly. Morgan did not contact him to inform him about what was about to happen. Edison was shocked, hurt, and furious, his pride deeply wounded. His name had been stripped from the company he founded, he had been pushed aside, and he had been treated shabbily. He was understandably bitter that his company and all his patents had been turned over to the competition.

The merger was finalized in April. At this point, Edison was effectively out of the electrical business. Although he had been named a member of the Board of Directors of General Electric, he never attended a board meeting, vowing that he would never sit on a board he did not control. To save face, Edison told the world he already was on to bigger and better things, noting that "I cannot waste my time over electric-lighting matters, for they are old."[15]

In twelve short years, Edison had made one of the most important inventions in history, commercialized it, and been a driving factor in the growth of the industry it created—and then had his name erased out of corporate existence. The merger forming General Electric was the second-largest ever completed up to that time. The electrical industry had consolidated from fifteen companies five years earlier to two: General Electric and Westinghouse. Even though GE now controlled 75 percent of the market, GE and Westinghouse would continue to compete fiercely. The battle of currents, however, was effectively over. Both companies were selling AC and DC systems.

Coffin served as General Electric's president until 1912 and board chairman until 1922. By the time he retired, he had become one of the wealthiest men in the world. Coffin's role in

---

[15] *Ibid.*, p.242.

developing technology and corporate leaders led management guru Jim Collins, author of the best-selling book *Good to Great,* to name Coffin the greatest CEO of all time. A profile of Coffin posted on GE's corporate website gushes, "Customers and competitors knew him as both the outstanding statesman and the outstanding salesman of the electrical manufacturing industry." It further states that

> Coffin's associates ... knew him as a gracious gentleman and delightful companion. He never ordered one of them to do anything, preferring to rely on his powers of suggestion. In his turn, he graciously sought and welcomed suggestions from those around him and then decisively made up his own mind on key questions.[16]

The lofty image of Coffin promulgated today is in stark contrast with his reputation in the 1880s and 1890s as a fast-talking, scheming, ruthless, unscrupulous businessman.

In May 1892, on the heels of the formation of General Electric, AC was about to receive a tremendous boost in the public eye when Westinghouse won the contract to light the 1893 World's Columbian Exposition in Chicago. The installed system demonstrated the safety, efficiency, and reliability of a mammoth, fully integrated AC system. Each night, ninety thousand light bulbs came aglow to dazzle visitors. The system also powered motors that drove all kinds of equipment and exhibits, including the Exhibition's trademark Ferris Wheel. More than 27 million people attended the exposition during its six-month run.

---

[16] Quotations from a profile of Coffin retrieved from General Electric's corporate website on May, 17, 2018. https://www.ge.com/about-us/leadership/profiles/charles-coffin.

The final, highly visible victory for AC was the construction of an AC system to harness Niagara Falls to generate electricity and transmit it 26 miles to Buffalo. In 1893, Westinghouse was awarded a contract to build three (later expanded to ten) 5,000-horsepower generators. Three-phase transmission line and transformer contracts were awarded to General Electric. Westinghouse's successful Columbian Exposition installation was a determining factor in its huge contract award. The generators, five times more powerful than any previously demonstrated, were installed at Niagara Falls in an elegant powerhouse (dubbed The Cathedral of Power), designed by the famous New York architect, Stanford White. In 1895, the generators began supplying electricity to nearby power-guzzling Alcoa and Carborundum plants. Late in 1896, electricity was transmitted to Buffalo and began powering the city's streetcars. During the years to follow, the system continued to be expanded and improved.

None of this was accomplished without strife. Coffin continued his attempts to crush Westinghouse during the Columbian Exposition and the Niagara project. GE refused to sell light bulbs to Westinghouse for use in fulfilling its Exhibition contract, citing Edison's patents. When, to Coffin's complete surprise, Westinghouse individually invented a new "stopper" bulb to circumvent those patents, GE surreptitiously attempted to obtain a court restraining order to shut down production of stopper bulbs, claiming they also infringed. Any shutdown would have made it impossible for Westinghouse to fulfill its contract. Westinghouse, personally learning by chance that a GE lawyer was on his way from New York to Pittsburgh to file for a restraining order, had his attorney simultaneously show up in court unannounced. As a result, the judge denied the restraining order and soon thereafter ruled there was no infringement.

In May 1893, Westinghouse learned that many of the company's blueprints and documents about pricing, labor costs, and other privileged information pertaining to the Columbian Exposition and the Niagara project bidding were at GE's Lynn, Massachusetts, plant. He immediately obtained a search warrant, and GE was caught red-handed.

Later, Coffin and Morgan attempted an unsuccessful hostile takeover of Westinghouse by engineering a market crash in Westinghouse stock. Morgan subsequently finally quit trying to merge Westinghouse into General Electric to create his envisioned electrical market monopoly and instead, in 1896, negotiated a patent-sharing agreement between the companies. This saved immense patent litigation expenses being incurred by the companies and left an ongoing duopoly in the electrical market.

The Columbian Exposition and Niagara Falls AC systems were significant. The Chicago installation demonstrated a universal system capability that moved the electrical industry from an era of electric light to an era of electric light and power. Such universal systems could supply energy for incandescent lights, arc lights, DC and AC motors, and other uses from a common transmission line or ring fed by centralized large-scale generators. With this capability, a utility company did not have to scrap its existing distribution circuits. Niagara showcased a system capable of delivering large-scale power for a wide range of purposes in addition to lighting at substantial distances from where the electricity was generated.

The advent of AC systems enabled the major role hydroelectricity would play in years to come, with larger and larger dams generating ever increasing amounts of electricity for distribution to usage points further and further away. The economies of scale of centralized generating facilities made electricity increasingly affordable. The demand for electricity skyrocketed.

Fig. 3.3 George Westinghouse
Alternating Current Pioneer

Fig. 3.4. Henry Villard
President of Edison General Electric

Fig. 3.5. Charles Coffin
First President of General Electric

Fig. 3.6. J.P. Morgan
Powerful Financier and Banker

Fig. 3.7. Telephone, Telegraph, and Power Lines Over New York Streets

Fig. 3.8. Nikola Tesla
Inventor of the Brushless Polyphase AC Motor

Fig. 3.9. Columbian Exposition Aglow

Fig. 3.10. Niagara Falls Powerhouse

Chapter Four

Sammy

Once while they were in Chicago, Edison turned to his most-trusted lieutenant and said, "You know, Sammy, this is one of the best cities in the world for our line of business."[1]

The lieutenant was Samuel Insull, and he would prove the observation to be absolutely accurate. It was from Chicago that Insull led the way. From a tiny base in the Second City, he quickly built Commonwealth Edison into the country's leading electric utility. He pioneered system standardization and electric grids, developed rate structures and highly innovative programs to both stimulate demand for electricity and level the load on electrical grids, continuously drove down electric rates, prodded companies such as General Electric and Westinghouse to make breakthroughs in equipment capabilities, turned electric utilities into regulated monopolies, created unique financing vehicles to meet the industry's mammoth capital requirements,

---

[1] Forrest McDonald, *Insull*, Chicago: University of Chicago Press, 1962, p. 53.

and showed that rural electrification could be achieved profitably. He was the father of the abundant, standardized, cheap, and dependable supply of electricity that we all take for granted now.

Despite his accomplishments, not everything in life was rosy for Insull. He went from being a powerful, highly revered, wealthy, and philanthropic business magnate during the 1920s to becoming a broken, bankrupt, despised scapegoat for the Great Depression in the early 1930s.

Insull was born in London in 1859. Despite his lower-middle-class roots, he was able to attend a good private school at Oxford until, shortly before his fifteenth birthday and upon a change in the family's financial situation and major disagreements with his father, he found a job as an office boy with a firm of auctioneers.

There, at the urging of the firm's head clerk, he taught himself shorthand and supplemented his income with evening jobs as a stenographer. This led to a regular evening job as stenographer to bon vivant Thomas G. Bowles, editor of the weekly *Vanity Fair* magazine and future member of Parliament. *Vanity Fair* included articles about such things as fashion, political intrigue, theater, books, social events, and high-society scandals. Determined to advance himself and thereby obtain respectability and success, Insull undertook, under the guidance of Bowles, serious and extensive general self-education in bookkeeping and business, political economy, history, and literature. His time was precious; he typically slept four hours a night. He focused, he memorized everything, he systematized information. He did allow himself several diversions. Opera was a passion that would remain with him for the rest of his life. He joined and became secretary of a literary society.

It was there, when asked to give a presentation on self-help, that he discovered an article about Thomas Edison. Instantly, Edison became Insull's hero. He read everything he could find about the inventor.

Early in 1879, Insull responded to a newspaper help-wanted advertisement and obtained a position as secretary to an American bank's London resident director, Colonel George Edward Gouraud. Only after beginning this new assignment did Insull learn what he had stumbled into: Gouraud's main activity was acting as Edison's European representative. Gouraud made Insull the bank office's head bookkeeper and introduced him to corporate finance. He also gave Insull the opportunity to learn all he could about Edison's affairs. Insull did so vigorously.

That fall, Edison's chief engineer, Edward Johnson, came to London to supervise a telephone exchange installation while Edison was conducting his first successful tests of long-lasting incandescent light bulbs back in New Jersey. Johnson was amazed to find that Insull knew more about Edison's European affairs than either Gouraud or Edison himself.

Stimulated by Johnson, Insull quickly learned everything he possibly could about the technical aspects of Edison's work. Before long, Johnson hatched a plan. He knew that Edison was highly organized and systematic in pursuit of his inventions but ramshackle and chaotic in his business and personal affairs. Insull would be the perfect private secretary and business nursemaid for Edison. Johnson would start extolling Insull's capabilities and wait for an opportunity to have him come to America in the envisioned role.

The wait was a short one. In January 1881, Edison's private secretary abruptly quit. Johnson cabled Insull, who immediately booked passage to New York. Johnson met Insull when his ship docked on February 28 and whisked him to meet Edison. At that first meeting, neither man was what the other expected. Each said to himself, "My God! He's so young!" Insull was twenty-one and looked sixteen. He was short, skinny, and fuzzy-cheeked. His dress was British-proper, his manner overly formal. His Cockney accent was so strong that Edison could barely understand him. Insull stood mouth agape staring at his idol. Edison, on the other hand, had just turned thirty-four. He looked disheveled, wearing a seedy Prince Albert coat and waistcoat, rumpled black trousers, a large sombrero, a rough brown overcoat, a dirty white shirt, and a white silk handkerchief carelessly knotted around his neck. He was ill-shaven. His manner was totally casual. Insull had difficulty understanding his thick midwestern accent. Nonetheless, they immediately started to work together through the night.

Edison was about to start building his Pearl Street power station, had already spent the funding he had received from investors, and wanted to use his European telephone securities to raise cash to proceed. By 4:00 a.m., Insull had gone through Edison's books. He had devised a complete schedule of Edison's European patent rights and outlined what additional money could be borrowed against them to finance the Pearl Street construction. Edison was amazed—and hooked on Insull's ingenuity and thoroughness.[2]

Insull became totally enmeshed in Edison's personal and business life. He became corporate secretary of all Edison companies. He prepared the business plan for the Pearl Street power station. He met the myriad fascinating people whom Edison attracted. In Spring 1883, after Pearl Street had been operational

---

[2] *Ibid.*, pp. 20–22.

for more than six months without meaningful, additional central plant sales (while several hundred isolated, special-purpose Edison systems had become operational), Edison charged Insull with the task of selling and building central stations. Then, when Edison ran out of manufacturing space and had labor issues in New York City, Insull located a facility about 165 miles north, in Schenectady, and set up and managed manufacturing operations there. With J.P. Morgan and his allies refusing to invest further, Insull also continually had to find ways to obtain bank financing for the manufacturing operations' growth and Edison's research. As a result, Insull developed a strong antipathy for New York bankers, especially those affiliated with Morgan.

By the time Morgan forced the merger of Edison General Electric into Thomson-Houston to form General Electric in April 1892, Insull, at the age of thirty-two, was, in effect, Edison General Electric's chief operating officer. Embittered by how the formation of General Electric had been handled by Morgan, upset that he had been named second vice president rather than president of the new company, and unwilling to work for Charles Coffin, Insull resigned. At the time he left General Electric, six thousand employees were reporting to him.

Coincidentally, Insull had received letters from two directors of the Chicago Edison Company (after 1907, the Commonwealth Edison Company) asking him to help find a president for the company. Remembering what Edison had said to him about Chicago's potential and wanting to be able to run a business in his own way, Insull offered himself as a candidate. The board of Chicago Edison ecstatically accepted his offer. The move to Chicago Edison would be quite a change for Insull. He would be going from a $50 million corporation to an $885,000 one. His salary would drop by two thirds. He would be leaving electrical manufacturing to enter the utilities field. And even though Chicago had more than a million inhabitants, there were fewer than five thousand users of electric lights.[3]

More than twenty companies were vying to provide electricity to users in the area. Nonetheless, during his General Electric farewell dinner, Insull brashly asserted that he would make Chicago Edison larger than General Electric itself.

Insull joined Chicago Edison full-time in July 1892. At the time, metropolitan areas such as Chicago were dotted with small, localized power plants serving nearby customers. There were no large-scale electrical utilities, and their technical and economic requirements were little understood. Insull would change that. Upon arriving in Chicago, he moved quickly on multiple fronts.

First, he personally borrowed $250,000 from famous Chicago retailer Marshall Field to purchase Chicago Edison stock, thereby providing capital for his ambitious expansion plans. He set out to acquire competitors, with the goal of achieving a monopoly. This would give him access to many more potential customers geographically, allow him to achieve economies of scale, and let him control rates. He wanted to make electricity cheap enough to be available to the masses, so those economies of scale would be used to relentlessly drive down rates. He intuitively understood the concept of elasticity of demand: by lowering rates, the resulting increase in business would lead to greater profits.

Within eight months, he had acquired his two largest competitors. By the end of 1895, he had acquired seven more, with even more—and their service areas—to follow. Importantly, via his acquisitions, he obtained exclusive rights to use every major

[3] Chicago's population would continue to explode by about half a million people during each of the next four decades, reaching nearly 3.4 million in 1930.

kind of electrical equipment (e.g., generators, transformers) in his areas of operation.

Insull simultaneously acted to increase his central station capacity to be able to generate the electricity necessary to service the expanded customer base he envisioned. He knew that the larger the generating equipment, the lower the unit cost of capacity. He also knew that designing power plants to allow the addition of more capacity in the future was critical to controlling costs and response time as he expanded business. He ordered his engineers to design a central station to be located on Harrison Street adjacent to the Chicago River, near where the river splits into the North Branch and the South Branch. The location was chosen so that coal could be transported to the site by barge and so that river water could be used in condensing generators.

He then struck a deal to purchase two generators General Electric was exhibiting at the Columbian Exposition for delivery upon the closing of the Exhibition at the end of October 1893. These were the largest and most efficient generators in existence at the time. When the Harrison Street station became operational in 1894, it was, by a wide margin, the largest in the world. Insull would continue to prod manufacturers relentlessly to develop larger, more efficient equipment for him.

Perhaps even bigger challenges were determining how to attract more customers, what customers to focus on, and what to charge them. At first, Insull expanded his base by offering potential new, large customers whatever rate they might be willing to pay in exchange for a long-term supply agreement. This led to revenue and customer growth, but he still wasn't sure what the path to profitable growth should be. A return to England for the Christmas holidays in 1894 led him to the answer.

Arriving in the evening at the resort town of Brighton, he noticed that every shop, small or large, seemed to have electric lights blazing. Amazed, he contacted the local electric plant manager, who explained that the cost of producing electric light had two components, fixed costs and operating costs (such as the cost of coal to fuel generators), each varying from customer to customer. It was necessary to invest in equipment to be ready to meet a customer's peak demand. The customer's service charge should be a combination of a fee related to his peak demand and a fee related to his usage. A utility's profits depended on what Insull came to dub the load factor—i.e., the percentage of time the plant investment was in use.

Stimulating customer usage at off-peak times of the day was particularly beneficial to keep plant investment in check. Because all customers used their equipment or lights only a fraction of the time and not all at exactly the same time, a single investment could supply multiple customers—and two customers always could be supplied more cheaply than one.

Insull concluded that the goal should be to service as many customers as possible, that utilities were economically superior to isolated plants, and that monopolies were the route to lowest rates for users of electricity. He directed his sales force to offer service to any prospective customer for less than he currently was paying for gas lighting or electric service. He introduced tier pricing that gave lower pricing for more usage and further discounts if electricity was used at off-peak hours. He metered customers' usage and established a statistics department to gather detailed data about costs and customer utilization.

With the new system, many homeowners found their bills lowered by 32 percent within a year. Chicago Edison's business boomed as a result of these sales efforts and new rate structures.

In Insull's first six years, production increased by a factor of eight to an annual rate of 26 million kilowatt-hours.[4]

To increase load factor, Insull targeted daytime customers to offset high evening usage by lighting customers. Streetcar and electric-railway power demand peaked between 6:30 a.m. and 9:00 a.m.; manufacturers required most of their power between 8:00 a.m. and 5:00 p.m.; and electric-light usage peaked between 5:00 p.m. and 9:00 p.m. In addition to focusing on the daily load factor, Insull also introduced the concept of a diversity factor, seeking customers whose electricity demands occurred at different times of the year. For example, ice houses and brick yards needed more power in summer months, whereas lighting demand was strongest in the winter.

Furthermore, in 1902, Insull obtained his first major contract with a transportation company. Within a short time, he had contracts with all the streetcar and elevated railway companies in Chicago. By 1908, his traction load (streetcar, elevated railway, subway, and interurban lines) had risen to 65 percent of sales.

Acquisitions and the rapid increase in business expanded the geographic area to be serviced beyond the one-mile distance limit for the direct-current Harrison Street central station. Insull was well aware that embracing alternating current would allow him to reach customers farther away from his central station. He had been a participant with Edison in the war of currents with Westinghouse, had seen Westinghouse's AC system demonstration at the Chicago Exposition, and was following the progress of the Niagara Falls AC power project. He understood the relative danger issues of AC. He also was convinced that, due to economies of scale, serving as many customers as possible from

---

4 McDonald, *Insull*, p.103.

a single large central station was more profitable than doing so from multiple, geographically separated, smaller central stations.

Insull began to link the Harrison Street station with central station locations of utilities he was acquiring and the sites of commercial customers who had been generating their own electricity. The DC current generated at Harrison Street was converted to AC using newly invented rotary converters and was stepped up with transformers to higher voltage for transmission. Small, inefficient generating plants of acquired utilities were closed and repurposed into substations where incoming high-voltage AC current transmitted from Harrison Street was transformed to lower voltage and converted back to DC. The electricity then was distributed from the substation as if the substation were a power plant. By also using frequency converters as needed, Insull could power the variety of user loads already in place. Existing investments in electrical wiring and user apparatus could be preserved. Isolated generating plants of commercial customers similarly were decommissioned and serviced from Harrison Street. Over time, he worked with equipment manufacturers to drive them to settle on standard voltages and frequencies in order to simplify network requirements. Insull's Chicago model was adopted by electric utilities across the United States and eventually became the foundation for today's national grid.

All these efforts led to exponential growth for Chicago Edison. But Insull's success in increasing demand for electricity soon created a major challenge: He was outgrowing the ultimate capacity of the Harrison Street plant. He needed a new plant and larger generators to increase the efficiency and capacity of his plants.

There were two ways to produce electricity in the early years: hydropower and coal-fueled reciprocating steam engines. Chicagoland was flat and lacked the falling water and high-volume river flowrates needed for significant hydropower generation. Coal, on the other hand, was readily available. The Harrison Street station thus had employed reciprocating steam engines that were powered by coal. Reciprocating steam engines were inherently inefficient, large, bulky, noisy, and difficult to maintain. Via linkages, they converted the up-and-down motion of pistons to the rotary motion of a flywheel to drive an electric generator. The 4,000 kW generators already installed at Harrison Street by 1900 were as large as considered technically practical.

Engineers for years had thought steam turbines could solve the reciprocating-engine bottleneck. Steam turbines channel steam into a narrow space so that the steam's pressure on radial blades can turn an attached shaft. This maximizes energy efficiency and is the same principle used in hydropower turbines, but the design specifics vary due to the vastly different properties of steam and water. Steam turbines promised to be smaller and lighter than equivalent reciprocating engines and scalable.

The first practical turbine, a reaction turbine, was invented in 1884 by Englishman Charles A. Parsons, who quickly sold the rights to Westinghouse for further development. In 1896, American Charles Curtis patented an impulse steam turbine, the rights to which General Electric acquired in 1897. After extensive development, General Electric, late in 1901, demonstrated a 500 kW turbine-generator unit. In April 1902, Insull prodded highly reluctant GE President Charles Coffin to manufacture two much larger 5,000 kW units. The two men eventually settled on a joint venture to share the risk.

The first of the 5,000 kW turbines was tested successfully in March 1903 at Chicago Edison's new Fisk Street (now Cermak Avenue) power station. It soon was generating more power than had ever been produced by any reciprocating steam generator. In another eighteen months, the 5,000 kW units were replaced by new GE turbines twice as large. Within a decade, Insull would be demanding and installing 35,000 kW units to power Chicago.

At the turn of the past century, Chicago was a raw, rough-and-tumble city. Its government was notoriously corrupt (some things never change). If Insull were to create and maintain a monopoly there, he would have to deal head-on with the politicians. And so he did. Local politicians wasted no time before attempting to extort money from him.

Under existing Illinois law at the time, city councils could enfranchise public utilities for a period of up to twenty years. That would soon change. In 1897, a week after the enactment of new state legislation allowing city councils to grant franchises for up to fifty years (legislation which, interestingly, would survive for only a year), a group of Chicago politicians sent messengers to tell Insull that he would have to pay as much as a million dollars to preclude their creating a competitor with a fifty-year franchise covering the entire city. Insull flatly refused their extortion. The politicians increased the pressure by voting a fifty-year franchise to a dummy corporation they called Commonwealth Electric Company. Only then did they learn that Insull had outfoxed them. He possessed the exclusive geographic rights to purchase the electrical equipment of every American manufacturer. The politicians would never be able to buy the equipment necessary to turn their dummy corporation into an operating business. Commonwealth Electric and its fifty-year franchise soon were sold to Insull's Chicago Edison for $50,000. Insull later officially merged the two companies to become Commonwealth Edison.

Illinois politicians suddenly had gained respect for Insull. This was not enough for him. He never bribed politicians. Instead, he cultivated relationships with them and made them allies. He made regular, large contributions to all candidates in all important elections. He made lower-level jobs that were justifiable from a business perspective available for ward bosses to distribute to their constituents. As he increasingly became a prominent—indeed, legendary—businessman, he usually was able to use his substantial influence as a business leader to obtain what he wanted.

From his earliest days in Chicago, Insull had believed that electric utilities should be monopolies to allow the best and cheapest electrical service to be accessible to the most users. Not everyone agreed. Due to monopoly abuses in the railroad industry and people's natural distrust of monopolies, progressive reformers were advocating that electric utilities be owned by municipalities. They argued that because cities had no stockholders demanding dividend payments or other investment returns, they theoretically could pass on savings directly to residents. Insull knew this would not happen in politically corrupt Chicago where politicians would raid the till.

For Commonwealth Edison to continue to thrive as a monopoly, Insull was convinced that certain things were essential. First and foremost, he had to continue to drive down rates so that consumers would not feel gouged—thus his emphasis on technology and economies of scale. Rates for commercial and residential customers, which were 20 cents/kWh when Insull arrived in Chicago, had dropped to 9 cents in 1908, and to 7 cents in 1912,[5] and would continue their decline due to his efforts. Other utility operators around the country were awed by this achievement.

---

5 Commonwealth Edison 1912 Annual Report.

It also was essential that the company be viewed positively by its workers, its customers, the community, and politicians. Insull published detailed annual financial reports for years before this became standard practice so that the public would know that his businesses were transparent, honest, and fair. He made sure that ComEd would be considered the best place to work in Chicago. The company's standard work week was forty-four hours—in an era when most other businesses had sixty- to seventy-hour weeks. Free medical benefits, profit-sharing, a retirement plan, and more were offered. Employees were strongly encouraged to participate in some form of community service activity. ComEd contributed conspicuously to major Chicago charitable and civic organizations. And as time passed and his empire expanded, Insull even began allowing employee associations to acquire stock in his companies at well below market price. Such treatment of employees was virtually unheard of at the time, and employees let their friends and neighbors know how well they were treated.

By 1898, Insull had conquered Chicago. He had received a fifty-year exclusive franchise within the city, and he and his company were highly regarded. Nonetheless, there was the constant need to deal with other municipalities in which he had franchises, the ever-present specter of municipal utility ownership, and the web of political corruption in cities everywhere. Because he firmly believed that statewide public utility regulation was a far better alternative to local politics and to municipalization, it was time to act.

That year, he had been elected president of the National Electric Light Association. Speaking at NELA's convention, he proclaimed the need for public regulation of privately owned utilities. He believed that to secure franchise protection, it was worth giving authorities control over rates. This *quid pro quo*—

i.e., exclusive franchises for cost-based rate maximums—would lower interest costs (a huge cost item for public utilities), thus lowering rates. Customers and investors would place greater trust in regulated utilities whose profit margin was negotiated in public view. Most of his fellow electric utility heads were opposed (many adamantly opposed) when Insull gave his speech. But he won them over in the years to come. With his skilled efforts as the industry's acknowledged leader, state after state implemented formal cost-of-service regulation for electricity. By 1914, forty-five of the country's forty-eight states had created state-level commissions to oversee electric utilities.

With his fifty-year exclusive franchise firmly in place in Chicago, Insull began turning his attention to Chicago's suburbs. He realized that weaving the city and suburbs into a single metropolitan power web would result in significant savings for consumers and profits for investors. He had been laying the groundwork for this move for years. In 1902, he had bought utility companies in Evanston, Highland Park, and Waukegan to form the North Shore Electric Company. The 1907 interconnection of high-voltage lines between Commonwealth Edison and the NSEC initiated an integrated metropolitan power network, creating a unified community of energy consumers on an unprecedented scale. The diversity of demand between the city and the suburbs was great enough to significantly reduce the unit cost of electricity for both groups of customers. Insull continued to repeat the process of acquisition, modernization, and expansion to establish a virtual monopoly of electrical services over the entire Chicago region.

While he was expanding his electrical web in the Chicago region, Insull decided to experiment in rural electrification. No one had believed this could be done profitably. Insull thought otherwise. He could string together a number of small villages with a high-voltage transmission line from a central source and either create a substation at each village or utilize existing equipment there as a substation connecting point. The high initial investment in transmission lines would be offset with every customer added, and the overall system's diversity factor would be improved since rural peak power usage differed from urban patterns (e.g., midsummer in rural areas versus midwinter in cities). His experiments to this end were surprisingly successful. Insull was the first to demonstrate that systematized service was technically and financially feasible in large rural areas. He began to urge other utility operators to embrace rural electrification. To his surprise, when he met in 1912 with Secretary of Agriculture James Wilson to seek federal government participation in a collaborative, widespread program of rural electrification, he encountered total disinterest. Nevertheless, he continued his crusade undeterred.

Insull's expansion throughout the entire Chicago region was the genesis of two elements that would propel electricity nationwide into the very fabric of American life: today's national electricity network and rural electrification. Insull continued to expand his sphere of influence all over the middle western states and beyond. By the late 1920s, his companies included 300 steam plants and 200 hydroelectric plants, served more than 4 million customers in 5,000 towns in 32 states, and produced about one eighth of the electricity consumed in the United States.[6]

Insull showed an astounding knack for public relations and marketing before the rest of the world even knew what the terms meant. He made public appearances whenever possible to tout the virtues of electricity and what it (and, of course, ComEd) was doing to improve people's lives. He started an advertising

[6] McDonald, *Insull*, p. 275.

department in 1901 and soon grew it into a public relations department. *Electric City*, a free magazine published by ComEd, was distributed in stores throughout Chicago.

He continually used creative sales promotions to boost electricity usage. For example, his sales force offered to wire six lighting outlets free of charge in new or older houses to induce homeowners to install electric service. Salesmen went to commercial strips outside the city center, rented a store, and converted it into a model of lighting with brilliant window displays, appealing advertising signs, and ornamental streetlamps. They then offered neighboring merchants special deals on similar fixtures. In one memorable sales promotion, electric clothes irons were sold door-to-door offering a free trial period followed by small monthly payments.

Later, Insull foresaw how popular broadcast radio would become, the potential it had for attracting new users of electricity, and how useful it could be as an advertising medium. In 1921, Westinghouse and Commonwealth Edison jointly established Chicago's first radio station, KYW, and soon thereafter ComEd began promoting sales of electric home radios. Between 1923 and 1930, more than 60 percent of Chicago families purchased sets.

In addition to being a visionary marketing expert, Insull was an adept financier. He never could have built his empire without being financially savvy. The electrical business was, and is, highly capital intensive. Insull eventually issued over half a billion dollars in bonds backed by the assets of Commonwealth Edison. His bonds included a variety of novel features which he devised, such as depreciation reserves. After he began selling stock to employees, other individuals clamored to invest, and Insull realized that selling company stock and bonds in small amounts to the general public was an untapped source of capital. Such retail sales were unheard of then. By 1930, a million people were shareholders of Insull companies.

Insull had formed his first holding company, Middle West Utilities Company, in 1912 as the primary vehicle for financing the acquisition of power companies outside the Chicago metropolitan area. The concept was not a new one. Similar holding companies already were being formed on the East Coast, driven by beliefs that financing could be better obtained by the holding company than by the captive operating companies, that the holding company could provide valuable management services and enhanced technology, and that the holding company could provide an integrated regional perspective. The capital structure of Middle West was deliberately designed to give Insull absolute control over a geographically diverse array of operating companies with a disproportionately small initial investment. Middle West eventually would have several hundred subsidiaries.

By the 1920s, Insull was indisputably the best known and most powerful figure within the utilities industry. As his biographer observed:

> In the hero-worshipping postwar decade, Insull became the Babe Ruth, the Jack Dempsey, the Red Grange of the business world. The people—butchers, bakers, candlestick makers—who invested their all in his stocks fairly idolized him, and even titans viewed him with awe. He measured up to America's image of itself: a rich, powerful, self-made giant, ruthless in smashing enemies, generous and soft-hearted in dealing with the weak. His doings, small and large, became a spectator sport, and they were reported and followed accordingly.[7]

---

[7] *Ibid.*, p. 237.

He ruled Illinois. He was a frequent guest at the White House. He mingled with European aristocracy. He was decorated by foreign governments, awarded honorary degrees by American and foreign universities, and honored by professional societies. He appeared twice on the cover of *TIME Magazine*.

All of this worked well for Insull as long as the utility industry continued to grow and underlying properties of highly leveraged holding companies continued to produce sufficient cash to service debt. In the 1920s, everyone believed demand for electricity could only continue to increase. No one foresaw the coming Depression.

Also, the utility industry was consolidating at a rapid pace as industry leaders envisioned national monopolies and were driven by ego and hunger for ever-increasing personal power. Insull and other industry leaders found themselves bidding-up to unrealistic levels the prices of acquisitions. His principal holding company piled layer upon layer of companies into its increasingly incomprehensible, pyramided structure. It became relatively easy to manipulate valuations. What had started as a sensible, straightforward financial structure evolved into a root cause of Insull's sudden, crushing reversal of fortunes early in the 1930s.

Befitting his stature in Chicago, Insull did big civic and charitable things. He had loved opera since his days as a poor, working London teenager, and regularly attended performances after moving to Chicago. In 1922, he became the Chicago Civic Opera's president and principal benefactor. He believed the opera should be a public asset on par with the other great operas of the world. To give the opera a fitting home artistically and to create a perpetual financial reserve large enough to support the company, he proposed a mixed-use building with a state-of-the-art performance facility and a high-rise office tower above the new opera house. Rent paid by office tenants would finance the opera. Insull masterminded the entire project and oversaw all the details of the building's $23.4 million construction (about $335 million in today's dollars). The gala opening night was days after the devastating 1929 stock market crash that signaled the beginning of the Great Depression. The new facility was dazzling, but what Chicagoans talked about most was the building's western elevation. It clearly was a throne. The uncrowned king of Illinois had built a forty-five-story monument to himself.

But trouble was brewing in his kingdom even as Insull basked in the glory of his triumphant Opera House opening. The first sign of trouble had come in early 1928, when he discovered that a financial buccaneer had stealthily been purchasing large blocks of stock of Insull companies. Until then, Insull's control had been assured by having fragmented ownership among many small, loyal shareholders. By mid-Summer, the stealth purchaser's holdings were considerably larger than Insull's own, and Insull became convinced that a control raid was imminent. To guard against this, he formed two tightly controlled investment trusts with interlocking ownership and directorships that sat over six layers of his other holding and operating companies.

Given the deep antipathy for New York bankers that he had developed in his days with Edison, Insull had for years assiduously avoided the New York financial community. He did financings in Chicago whenever possible. Otherwise, he looked to London. He was vocal about his disdain for New York, and that made for enemies keen on revenge.

That revenge would soon come. In mid-1930, the buccaneer investor offered to sell his holdings in Insull's companies back to him for $56 million. To maintain control, Insull agreed—only to find to his surprise that, despite earlier assurances, he could not raise the money required for the purchase from his usual sources in Chicago and London. For the first time since leaving General Electric in 1892, he had to turn to the detested J.P. Morgan forces in New York, for loans totaling $20 million. The only suitable collateral he could offer for the loans to his two investment trusts was stock from their portfolios.

If the market in Insull securities were to drop, the investment trusts would have to put up more collateral against the loans. If the market were driven down far enough, the New Yorkers would hold the entire portfolios of the trusts as collateral and thus control of Insull's entire empire. Morgan proceeded to methodically manipulate the markets for Insull stocks to do exactly that.

By December, the entire combined portfolios of the trusts were in the hands of bank creditors. Insull frantically tried to salvage the situation—borrowing to the limit of his personal credit on behalf of his companies, moving money from company to company to shore them up, drastically reducing expenses. It was futile; the House of Morgan was in control. Morgan's final tactic was to cast Insull in the role of villain largely responsible for the Depression and the plight of millions of unemployed Americans so that Morgan forces could then take over and rescue the masses from this terrible scoundrel.

The die was cast. In June 1932, the Morgan bankers forced Insull's resignation. Unemployed, broken, and bankrupt, he traveled with his wife to Paris to avoid the spectacle of scandal that was unfolding in Chicago. But this attempt to live in anonymity was short-lived. Franklin D. Roosevelt was campaigning for the Presidency and heaped scorn in stump speeches on the "power trust" and its figurehead Insull. Criminal charges quickly were brought against him, making him a public spectacle. He was acquitted of all charges in three sensationalized trials. His wife and he then lived in obscurity in Paris until he died in 1938 of a massive heart attack while waiting for a train in a Paris subway station.

How ironic it all was. Embittered by treatment at the hands of J.P. Morgan, Insull had gone to Chicago early in his career to build the country's premier electric utility. He had turned his back on New York bankers for nearly forty years before being brought down by the House of Morgan. The fatal downfall of his entire empire was the result of a $20 million desperation loan from Morgan—only about three years after Insull personally had raised $23.4 million just to build an opera house for Chicago. He ended his life deposed, despised, and destitute. He has largely been written out of history.

The downfalls of Insull and the power trusts were precursors to major regulatory changes in the 1930s. These included the Securities Act of 1933 to provide full and fair disclosure regarding securities sales, the formation of the Securities and Exchange Commission in 1934, and the Public Utility Holding Company Act of 1935. The general assumption behind this New Deal regulation was that Insull and other high-profile promoters and utility magnates had swindled the American people and that big business could not effectively be regulated by the states. These changes dramatically restructured the electric industry and marked the end of its entrepreneurial days.

Fig. 4.1. Samuel Insull Shortly After Becoming Edison's Private Secretary in 1881

Fig. 4.2. Insull on the Cover of TIME Magazine
November 29, 1926; November 4, 1929; May 14, 1934

Fig. 4.3. Chicago: The Electric City *Cover, August 1904*

Fig. 4.4. A Memorable Electric Flatiron Mass Marketing Scheme

Fig. 4.5. Listening to Radio in the 1920s

Fig. 4.6. Chicago Civic Opera Building
Insull's Throne Facing West Away from New York

Chapter Five

Dams, Dams, Dams

Water, essential for all forms of life on the planet, covers 71 percent of earth's surface. Harnessed or unharnessed, it is a powerful force. We have only to fly over the Grand Canyon or drive through the Dakota Badlands to marvel at what the flow of water has done over the ages to gouge and shape the earth's surface.

The watershed system of the United States indeed is a wondrous thing. The Mississippi Basin drains more than 40 percent of the country's land area, from the Appalachians to the Rockies. Among the Mississippi River's major tributaries are the Ohio, Tennessee, Missouri, Arkansas, and Red rivers. The St. Lawrence River drains the Great Lakes. In the Southwest, the Colorado River traverses seven states and Mexico on its route to the Gulf of California, and the Rio Grande forms part of the nation's southern boundary. The Columbia River gathers water from the Rocky Mountains and the Cascades, and the Sacramento and San Joaquin rivers collect water from the Sierra

Nevada range, linking inland valleys to the Pacific Ocean. Altogether, there are an astonishing 1.2 million miles of rivers and streams in the United States—the equivalent of five trips from earth to the moon.[1]

Man has been attempting to change the natural flow of water for eons to control flooding; supply water for agriculture, industries, and households; assist river navigation; turn water power into mechanical power; and create recreational opportunities. Dams have been integral to those attempts.

A dam simply is a barrier that stops, restricts, or regulates the flow of water, raising the water level behind it. Dams have been created in one form or another since the beginning of time. Mother Nature has created her share. Volcanic dams are formed when flowing lava encounters a stream or lake outlet. Glacial activity, earthquakes, and landslides can form natural dams. Debris swept downstream by flood waters can become entangled and eventually create a significant dam. Some animals also are dam builders. The beaver, for example, is a masterful dam builder, harvesting trees to create one or more dams to provide still, deep water to protect its lodge against predators and to float food and building material.

The earliest manmade dams for which remains exist were built around 3,000 BCE as part of a water supply system for the town of Jawa in Jordan. Remains there are of three dams spanning a dry riverbed which flooded sporadically during winter months. Two of the three were deflection dams to channel water into a number of reservoirs. The third was a reservoir dam that completely blocked the flow of flood waters.

---

1 US Environmental Protection Agency, *National Rivers and Streams Assessment 2008-2009: A Collaborative Survey*, Washington, DC, 2016, p. 16.

Remains also have been found of another good-sized dam, Sadd el-Kafara, that was under construction near Cairo, Egypt, for flood control circa 2,600 BCE when a flood partially destroyed it. The dam had stone-block exterior walls encasing a core of earth and rock fill.

By the late first millennium BCE, stone and earth dams had been built around the Mediterranean and in the Middle East, China, South Asia, and Central America. The Romans built dams throughout their Empire. In Europe, with rainfall ample and relatively well distributed throughout the year, dam construction before the Industrial Revolution was on a modest scale, restricted to forming water reservoirs for towns, driving watermills, protecting areas prone to flooding, and recovering water losses in navigation canals.

The history of converting the energy of flowing water into mechanical energy also is long. Waterwheels rimmed with buckets were used in ancient Egypt and Sumer to scoop water from rivers and canals. By the first century BCE, watermills were grinding corn in Rome. *The Domesday Book,* a comprehensive survey compiled under the orders of William the Conqueror, recorded 5,624 watermills in England in 1086.

Dam building came to the Americas with the conquistadors. In Mexico, they imitated the dams built by the Romans, Muslims, and Spanish Christians in their homeland to construct dams for irrigation.

Native American tribes were not prolific dam builders in what is now the United States. Although the Hohokam people were building sophisticated irrigation systems in the Southwest more than a thousand years ago, most tribes used rivers as highways and as sources of fish for food. They also saw rivers as living spiritual entities.

Early colonial settlers along the Atlantic coast of America brought dam-building knowledge with them from Europe and quickly put it to use in their settlements and towns as they ventured westward. The first major structure in a settlement usually was a church, followed by a dam. The dams plugged streams and set them to work turning gears to grind corn and to saw lumber, while also storing water. The oldest of these that still is in use was built in Scituate, Massachusetts (about 30 miles south of Boston), in 1640.[2] In the nineteenth century, dams controlled the rivers that powered the mills that produced goods such as flour and textiles. These dams were of earthen, timber, or masonry construction.

Falling water supplied a major portion of industrial power requirements in the United States throughout the 1800s, a far larger fraction relative to steam than in Europe. With plentiful water resources readily accessible in the northeast and north-central parts of the country, American manufacturers built waterpowered factories on a never-before-seen scale. Starting in the early 1820s, harnessing the power of Pawtucket Falls on the Merrimack River via a system of canals, waterpowered textile mills were built in what became the famous factory town of Lowell, Massachusetts. Lowell quickly became the country's largest industrial center, with a 5.6-mile-long canal system powering forty mills and a population that had swollen to 33,000 people by 1850. It now often is called the "Cradle of the American Industrial Revolution." Lowell's success was imitated extensively as similar waterpowered factories rapidly were built elsewhere.

---

[2] US Army Corps of Engineers, *National Inventory of Dams*, Washington, DC, January 2019.

Natural waterfalls facilitated dam building. Eastern rivers tended to fall gradually over relatively flat terrain. Consequently, eastern waterpower systems usually were characterized by low dam height. On streams where waterfalls could not be directly tapped, the dams used to raise the stream level to create falling water typically were 5 to 20 feet tall.

The situation differed west of the Rocky Mountains. Other than for irrigation and water-collection needs of the sparse population in the mountainous west, there was little demand for dam construction—until the California Gold Rush of 1848.

Miners needed power for hydraulic mining to wash gold-bearing sand and gravel from hillsides, for hoisting, and for ore crushing. Once trees near a mining operation were consumed, fuel for wood-burning steam engines was unavailable as a practical matter. Water power was a highly-sought-after alternative. Rainfall in the west was sparse, but falling-water pressure heights of hundreds of feet were available if mountain streams were dammed.

As the population of San Francisco burgeoned after the Gold Rush, and with the completion of the first transcontinental railroad twenty-one years later, the first dam in the world to be made entirely of concrete was built near San Francisco.[3] It was completed in 1890 in order to impound a reservoir to meet the city's water needs. The San Mateo Canyon Dam (also known as the Lower Crystal Springs Dam), located about 100 yards from the San Andreas earthquake fault line, was a 154-foot-high, double-curvature, arch-gravity dam with a sophisticated design. It was built of interlocking concrete blocks formed and poured in place. There was no cement industry in California at the time, so Portland cement was imported by sea from England

---

[3] Norman Smith, *A History of Dams*, London: Peter Davies, 1971, p. 210.

and mixed into concrete in accordance with exacting specifications. The dam survived the 1906 San Francisco earthquake undamaged and remains in operation today.

Double-curvature, arch-gravity dams are but one type of dam. There are myriad design options, though dam configurations fall into three basic categories: embankment, gravity, and arch (see Figure 5.1).

Figure 5.1: Basic Dam Configurations

- **Embankment or Fill Dams:** These dams can be either earth-filled or rock-filled. They rely on their heavy weight to restrain the force of the upstream water. They typically have a dense, waterproof core that prevents water from seeping through the structure. Alternatively, there might be an impervious upstream facing made of concrete, masonry, plastic membrane, timber, or other material. When embankment dams can be constructed from materials found on site or nearby, they frequently can be cheaper to build than concrete dams. They often are used where there are broad water courses. Oroville Dam, at 770 feet the tallest dam in the United States, is an earth-filled embankment dam on the Feather River in California.

- **Gravity Dams:** These dams generally are made of either concrete or masonry. They are held in place by earth's gravity, which pulls down on the mass of the dam to overcome the push of water against the upstream face. Gravity dams are the best-developed type of dam configuration. An example of this type of dam is the massive Grand Coulee Dam on the Columbia River in the State of Washington.
- **Arch Dams:** The arch dam is one of the most elegant civil-engineering structures. The dam is thin in the cross section compared with gravity dams and therefore uses less material. When viewed from above, it is curved so the arch faces the water and the bowl of the curve looks downstream. The forces imposed by upstream water are carried into side abutments. The arch dam is most suited to narrow canyons with steep walls composed of sound rock. Utah's 502-foot-tall Flaming Gorge Dam is an arch dam.

There are variations of gravity and arch dams, including arch gravity (which uses the weight of the dam to offset additional water forces), buttress, and multiple arch. Hoover Dam is an arch-gravity dam.

Arch dams are a relatively recently introduced type of dam structure. Bear Valley Dam, the first arch dam built in the United States, was the culmination of Frank E. Brown's dream of creating an irrigated agricultural colony in Redlands, California. After graduating from the Sheffield Scientific School at Yale University and taking a graduate course in civil engineering, Brown moved to southern California in 1877. He investigated sites high in the San Bernardino Mountains above Redlands for storing snow melt in a reservoir that could be tapped for irrigation water. The Bear Valley seemed ideal—and it had a narrow granite gap at its outlet where a dam could be built. Applying what he had learned at Yale, Brown used mathematical analysis to design a daring arch dam. Engineers who heard about the design asserted that the boldly thin dam would not hold and declared it "The Eighth Wonder of the World" when it did.[4] The masonry structure, completed in 1884, was 64 feet tall and had a crest length of 450 feet. The dam's thickness tapered from 22 feet at the base to an astoundingly thin 3 feet near the top. Yale awarded Brown a civil engineer's degree for his work on the project.

Since around 1900, concrete construction has been used more often than fill or masonry construction. Concrete possesses more reliable and predictable properties than other materials. It became even more attractive as dam-building material as its compressive strength was gradually increased during the first half of the twentieth century. Concrete also provides design flexibility. For example, gates or other outlet structures can be built directly into the concrete.

Varying types of concrete have been used for years. Ancient Romans used a material that is remarkably close to modern concrete to build many of their architectural marvels, such as the Colosseum and the Pantheon. Portland cement, invented in 1824, forms today's concrete when mixed with water and aggregates (rock and sand). Portland cement is a closely controlled chemical combination of calcium, silicon, aluminum, iron, and small amounts of other ingredients, to which gypsum is added to regulate the setting time of the concrete.

---

4 From inscription on California State Historical Landmark Number 725 (Old Bear Valley Dam) plaque.

The curing of mixed concrete is an exothermic reaction—that is, it produces heat. If a concrete slab is thick enough and the heat is not sufficiently controlled, the temperature can rise to the point where thermal stresses cause unwanted cracking. This phenomenon has made concrete construction a precision process.

Furthermore, concrete naturally has high compressive strength (strength when squeezed), but low tensile strength (strength when stretched). Iron-reinforced concrete construction first was introduced in 1853 to augment tensile strength. Techniques developed for using steel reinforcing bars (rebar) in concrete construction were essential to the eventual widespread use of concrete, partially to add structural strength and partially to offset the effects of heat generated during the curing process.

Until about 1850, dam design was completely empirical: Designers extrapolated from what had worked and what had failed. The Industrial Revolution stimulated developments in science and engineering, and the civil engineering specialty came into existence by the 1850s. Early civil engineers studied Sir Isaac Newton's laws of physics and other scientific theories and applied them to practical structures, including dams. By 1886, Cornell University's Civil Engineering Department was offering graduate-level courses in bridge, railroad, and dam engineering.

The first mathematical attempts at structural analysis for gravity dams were made in the 1850s. Major advances followed in understanding the relationship between the precise weight and profile of gravity dams and the thrust of water upon them. The pioneering work of François Zola (the father of famous writer Émile Zola) undertaken in France in the 1830s and 1840s on the theory underlying arch dam design also was continually refined and extended. By early in the 1900s, dam designers had an arsenal of scientific tools to facilitate and validate their work. This allowed them to push limits in ways previously not dared. These design tools were sound as far as they went, but more exact stress analysis techniques using equations of the theory of elasticity and experimental data continued to evolve during the first half of the twentieth century. The advent of computer analysis in more recent times greatly enhanced the civil engineer's design tool kit.

The high cost of building impound dams prompted ongoing innovations in dam building technology early in the twentieth century to reduce construction costs. Structural analysis was used to avoid overly bulky dam profiles. Newly devised design approaches required smaller volumes of material than traditional gravity-dam designs (for example, designs combining arch and gravity features). Reinforced-concrete construction used less skilled labor and less expensive material than traditional mortared cut-stone construction. And as dam projects continued to multiply in number, construction practices were refined and increasingly became more systematized.

Dam design and location also were driven by site topographic and geological conditions, access to and cost of construction materials, availability of construction labor, financial constraints, the experience of the design engineer, and the public image the project might convey.

No matter the design, however, dam construction was labor intensive, grueling work as crews fought nature and raced the clock before seasonal floods could wash the fruits of their labor downstream. Thousands of workers toiled on some of these projects, living in construction villages erected on site. Well into the twentieth century, the hand shovel remained the primary earthmoving machine, and construction materials were moved at the construction site by draft animals or human power. Whenever possible, specially constructed rail lines were used to transport materials to the construction location. This often was not practical for remote, mountainous sites.

The full-swing, 360° revolving steam shovel was developed in 1884, eventually becoming a workhorse for dam construction and easing the load for laborers and draft animals. Trucks and other motor-powered heavy equipment did not come into common use until the 1930s.

In these pre-OSHA days, dam construction was exceedingly dangerous work (the Occupational Safety and Health Administration was not formed until 1971). The prevailing working conditions would be unthinkable today. For example, early versions of hard hats were introduced just after World War I, but the first construction project requiring workers to wear hard hats was the building of Hoover Dam in 1931.

Dangerous working conditions weren't the only things that gave dam designers and builders pause. Two words—dam failure—always have given dam designers and operators chills. Dam failures, or dam bursts, are rare, but they can cause immense damage and loss of life when they do occur. The deadliest dam failure in US history happened in 1889, when the South Fork Dam near Johnstown, Pennsylvania, collapsed after several days of heavy rainfall. The resulting flood killed 2,208 people. In 1928, the St. Francis Dam collapsed near Los Angeles, about twelve hours after being inspected by the dam's designer. The massive flood wave swept 54 miles down to the Pacific Ocean and killed at least 450 people. More recently, the main and emergency spillways of California's Oroville Dam were damaged after seasonal rains in 2017, prompting the evacuation of more than 180,000 people living downstream.

Advances in dam design and building, along with the Age of Electricity (which was ushered in by Thomas Edison) and the westward expansion in the United States combined to stimulate a great era of dam building. The demand for power was insatiable in the early 1900s, and waterpower became an important solution. As the western areas of the country became populated, people moving there required water. The West developed through the construction of dams because it allowed the control of water for that development.

In the early days of hydroelectricity, as seen with Appleton's Edison system in 1882, electric system installations most often took advantage of existing dams suitably modified to drive generators. By the late 1890s, with universal AC electrical system capabilities demonstrated and equipment commercially available from General Electric, Westinghouse, and other component suppliers, new dams were being built optimized for electricity generation.

Hydroelectric power is simple to understand conceptually. As shown in Figure 5.2, water collected behind a dam is funneled into a pipe called a penstock and directed downward to a waterwheel or turbine. The force of the falling water on the turbine blades rotates the turbine and its main shaft. The other end of the main shaft is connected to a generator. Rotation of magnets attached to the shaft within the generator produces electricity. After water has passed the turbine blades, it continues through the tail race into the river downstream of the dam.

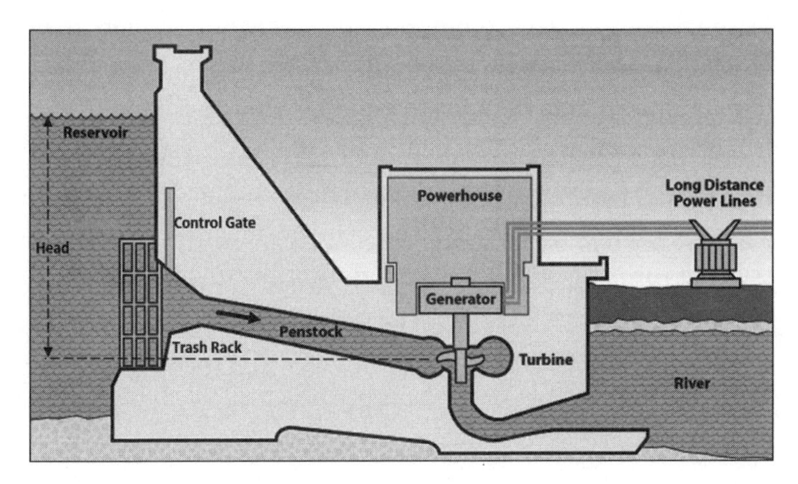

Figure 5.2: Hydroelectric Facility Schematic

The power transmitted to the turbine by the water rushing past its blades is a function of two factors: 1) the vertical distance between the surface of the water in the reservoir behind the dam and the turbine (usually referred to as the hydraulic head), and 2) the volumetric flow rate of the water dropping down through the penstock.[5]

Note that the entrance to the penstock is designed to be located far enough down the upstream dam face that it is always below the reservoir water surface, as river flow conditions can vary significantly. Control gates installed near the penstock entrance are used to control the rate of water flow down the penstock,

[5] The power P transmitted to the turbines can be expressed mathematically as $P = \rho g Q h$, where ρ is the density of water, g is the acceleration of gravity, Q is the volumetric flow rate of the water, and h is the hydraulic head. This equation explains the theoretical amount of power transmitted to the turbine by the water flowing down the penstock. Unfortunately, 100 percent of the theoretical amount of power is never converted to electricity. Turbine efficiencies, however, typically exceed 90 percent.

and trash racks prevent debris from entering the penstock.

Various types of turbines are used at hydropower facilities, with the type chosen being largely dependent on the plant's hydraulic head. The three most common types today are the Pelton turbine, the Francis turbine, and the Kaplan turbine.

There are two basic turbine categories: impulse turbines and reaction turbines. The water wheel is an early example of an impulse turbine. This predecessor to modern turbines was used for many years to power grain mills and factories, such as the textile factories in New England in the 1800s. Flowing water was directed into buckets set in the periphery of the water wheel. The impulse of the water stream hitting the buckets and the water's weight while in the bucket pushed on and turned the wheel around its central shaft, converting water power into mechanical power.

The Pelton wheel (or turbine), developed over the period of 1878–95 in the mining areas of the western United States, is a variation of the water wheel where one or more free jets of high-pressure water are discharged into an aerated space and then impinge onto specially designed curved blades around the periphery of the turbine's circular-disc rotor (i.e., the turbine's runner). It is akin to directing a fire hose at the rotor blades. No pressure change occurs at the blades, and the turbine does not require a housing for operation. Pelton impulse turbines are used for high-head sites.

Unlike impulse turbines such as Pelton turbines, reaction turbines develop power from the combined action of pressure and moving water. The runner is placed directly in the water stream flowing over the blades rather than striking each blade individually. Reaction turbines must either be encased to contain the water pressure or be fully submerged in the water flow. Reaction turbines typically are used for sites with lower head and higher flows than sites where impulse turbines are used.

The Francis reaction turbine was the first modern-day turbine. This reaction turbine, developed in 1849 by James B. Francis, an engineer in the textile-factory city of Lowell, Massachusetts, remains the most widely used water turbine today. It has a runner with nine or more fixed blades that direct water flow from the outer circumference toward the runner axis and downward. It is suited to a wide range of heads and flows.

The Kaplan reaction turbine, invented in 1913 by Austrian professor Viktor Kaplan but not commercially successful until 1924, is an evolution of the Francis turbine and was designed especially for low-head applications with high flow rates. The runner has three to six propeller-like blades whose pitch can be changed.

Water turbines can be large and heavy. Francis turbines have been installed that have 33-foot runner diameters and weigh more than 450 tons.

Unlike conventional impoundment hydroelectric dam facilities, diversion facilities funnel a portion of the flow of a river to the plant's turbine system. Pipes, tunnels, or canals create this flow diversion. A small dam often is built to create a head pond to ensure that enough water is fed to the turbines despite fluctuations in daily river flow. The output power of these run-of-the-river facilities depends heavily on fast-flowing water from the river. Because waterpower is not stored with run-of-the-river facilities as it is in impoundment dam facilities, large-scale diversion facilities must be installed on rivers that have consistent and predictable flow rates.

Pumped-storage hydroelectric facilities (see Figure 5.3) are variants of traditional impound-dam hydroelectric facilities.

They have two reservoirs at different elevations. In power-generation mode, water from the upper reservoir flows downward through the powerhouse to generate electricity. Water exiting the powerhouse flows into the lower reservoir rather than directly re-entering the river and flowing downstream. Turbines in the powerhouse are reversible. In off-peak times, excess electrical energy from the power grid drives the reversed turbines, which act as pumps to move water from the lower reservoir back uphill into the upper reservoir. Pumped-storage facilities often are called water batteries. They are used for electric power grid load balancing and typically can respond to load changes within seconds or minutes. Excess energy from intermittent sources (such as solar or wind) or continuous base-load sources (such as coal or nuclear) can be saved for periods of higher demand. The nation's first pumped-storage power station, the Rocky River Hydro Plant near New Milford, Connecticut, became operational in 1929. Pumped-storage facilities are used extensively today. They will continue to play an important role as the United States pursues its clean energy goals.

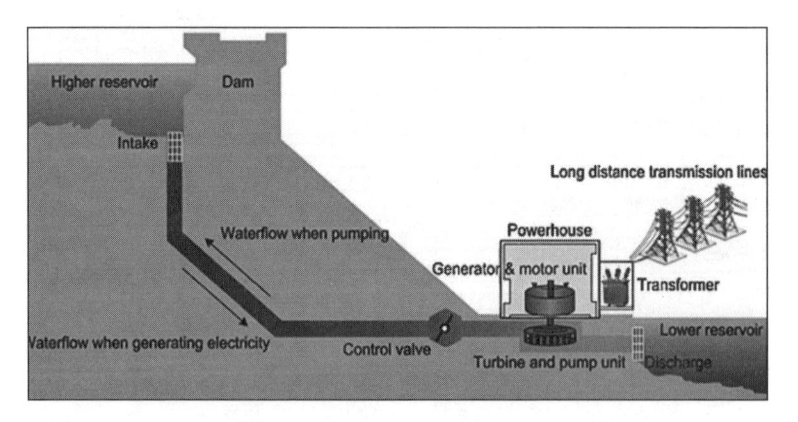

Figure 5.3: Pumped-Storage Hydroelectric Facility

In the early 1900s, hydroelectric power accounted for more than 40 percent of the United States' supply of electricity. In the 1940s, hydropower provided about one third of the country's total electrical energy. With increased development of other forms of electric power generation, the use of hydropower has slowly declined. Today it provides about 7 percent of America's electricity-generation capability.[6]

Because the installation of a hydroelectric facility requires the construction of a dam, reservoir, and powerhouse, relatively large capital investment is required. Nonetheless, hydroelectric facilities are cheap to maintain. Few employees are required to run the power plant. The necessary water is free. Hydroelectric power plants have a very long lifespan and require little maintenance.

Hydroelectricity from its earliest days was economically attractive. And there was a constant drive for larger and larger projects. Samuel Insull and other utility industry leaders knew that the larger the generating equipment, the lower the unit cost of capacity. Also, as investors considered new dam installations, they realized that, if the facility were to meet multiple objectives—such as irrigation plus electricity generation, flood control plus hydropower, or navigation plus electricity—the sale of the resulting electricity could produce a revenue stream to more than cover the overall capital investment.

As time passed in the first half of the twentieth century, record after record was broken for dam size, for electricity generated, and for area flooded. Historians compared American dams to Egyptian pyramids and Roman coliseums. Presidents spoke glowingly of dams as symbols of American engineering

---

[6] US Energy Information Administration, Washington, DC, *Electric Power Monthly*, February 2020.

prowess. People who visited dam sites gazed at them in absolute awe.

About fourteen thousand dams were built between 1900 and 1949 in the United States. There now are over 90,000 dams in the country, most of which are small and privately owned. The 3 percent of them that currently generate power provide about 39 percent of all renewable generation. Only 6,400 of them are more than 50 feet tall.[7] The tallest is Oroville Dam, which is 770 feet tall (equivalent to the length of 2.6 football fields). The largest source of hydroelectricity in America is Grand Coulee Dam on the Columbia River in Washington State, a concrete gravity dam that currently has an installed capacity of 6,809 megawatts. It is one of the largest structures ever built by humankind, containing enough concrete to build a highway from Seattle to Miami. It supplies enough power to continuously supply the needs of two cities the size of Seattle.

About 70 percent of the country's hydropower is produced in the West. The northwestern states of Washington, Oregon, Montana, Wyoming and Idaho generate approximately 50 percent of all hydroelectric output. The mountains are high in the region, and water is plentiful—ideal for hydropower production.

Another 20 percent of national hydroelectric output comes from the southwestern states of Colorado, Utah, Nevada, California, Arizona, and New Mexico. While their terrain is similar to that of their northern neighbors, their climate overall is drier. Although forty-eight states now have at least some hydropower installed, hydropower is not practical in many parts of the

[7] US Army Corps of Engineers, *National Inventory of Dams*, Washington, DC, 2020.

country. Hydropower generation works best in areas with ample water flow and mountainous or hilly terrain.

The golden era of dam building that captured America's imagination during the first half of the twentieth century has long passed. Hoover Dam, Grand Coulee Dam, and various Tennessee Valley Authority dams—together among America's largest sources of hydropower—were public works projects borne of the Great Depression. These projects were crucial toward saving, supporting, and building the US economy and for creating desperately needed jobs.

These and most other major, multipurpose dams built from the 1930s on were built with taxpayer money. Many helped power the nation through the intense demands of World War II and the subsequent economic expansion. They enabled rural electrification. They converted the Southwest from barren desert into arable cropland and livable cities. And they mitigated devastating river flooding throughout the country.

As the best dam sites were used, newer, larger, multipurpose dam projects became more and more expensive. Increasing regulatory requirements, as well as environmental and societal concerns, made project approval more difficult and time consuming. Furthermore, the introduction of electricity-generating plants powered by nuclear energy in the late 1950s began to divert attention and investment dollars away from hydroelectricity and coal-fired generation. By the late 1970s, budget considerations led to reductions in pork-barrel funding for federal dam projects. Given today's fiscal realities, as well as federal priorities and societal demand for developing other renewable energy sources such as solar and wind power, major new federal funding for water projects is unlikely. Various government agencies even have gone so far as to argue that "most of the good spots to locate dams and large hydro plants have already been taken."[8]

Despite the attention given to other forms of renewable energy, hydropower remains the country's cheapest electrical energy source, and coal its most expensive, when all costs are considered (see Figure 5.4).

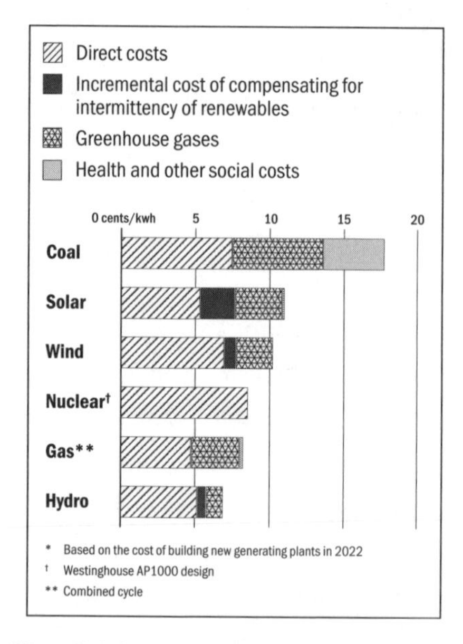

*Figure 5.4: Baseline Generation Cost by Source**

Our story, though, is not about the here and now; rather, it is about the amazing times during the golden era of dam building.

[8] See, e.g., US Department of the Interior, https://water.usgs.gov/edu/wuhy.html

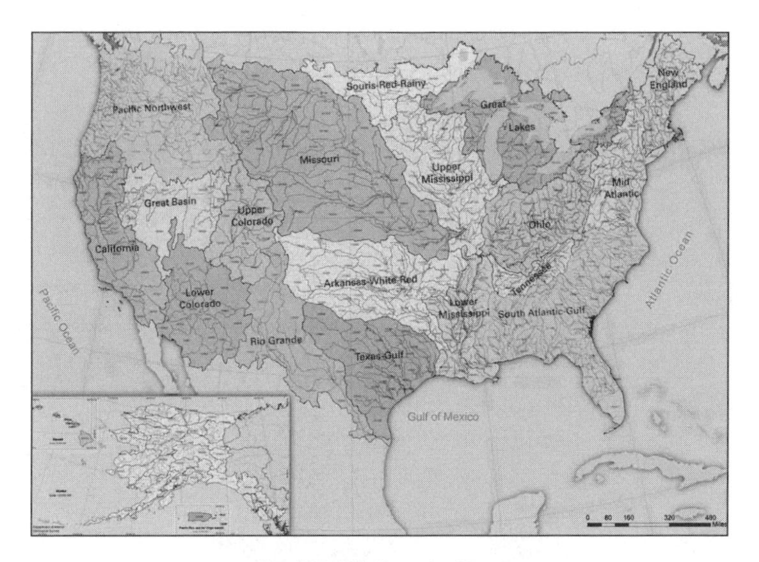

Fig. 5.5. US Watershed Regions

Fig. 5.6. Bear Valley Dam, America's First Arch Dam

Fig. 5.7. San Mateo Canyon Dam, the World's First Concrete Dam

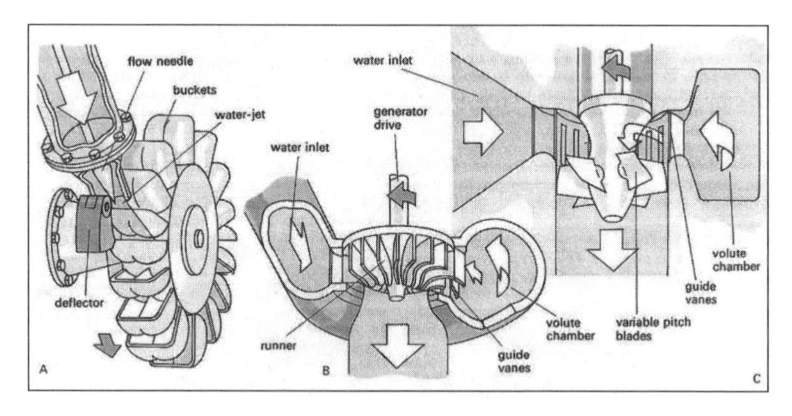

Fig. 5.8. The Three Main Types of Water Turbine
(A) the Pelton turbine (or wheel), (B) the Francis turbine, and (C) the Kaplan turbine

Fig. 5.9. Francis Turbine Runner, Grand Coulee Dam

Chapter Six

They Saw the Light

It was entrance examination day at the Massachusetts Institute of Technology in 1884, less than two years after Edison's pioneering Pearl Street power system had become operational in New York City. Two freshman engineering student candidates, Charles Stone and Edwin Webster, both from the Boston suburbs, met that day and quickly became inseparable friends. They would remain so during their years at MIT and, subsequently, as business partners for the rest of their lives. The two became such close friends at MIT that fellow students began referring to them collectively as Stone and Webster rather than as individuals.

"Charlie" Stone (1867–1941) and "Ted" Webster (1867–1950) were fascinated with all things electrical and decided to focus their studies on electrical engineering. It would turn out to be an incredibly wise choice—right time, right place. Edison's feats in lighting were the subject of continuing press adulation, the world saw itself being transformed by incandescent lighting, and these young men could sense the excitement around them—and the opportunities that electricity was sparking.

In 1884, in addition to the Pearl Street station, there were more than a dozen other Edison lighting system central stations in operation and roughly 400 additional operational, isolated plants. By August 1886, 58 Edison central stations and 702 isolated Edison plants were in operation.[1] Municipal systems were installed near Boston and MIT in Brockton and Fall River, Massachusetts, in 1883, and installations were to begin in Boston itself in 1886.

When Stone and Webster matriculated at MIT, the electrical engineering field was just beginning to emerge as an accepted discipline. MIT had established America's first four-year course in electrical engineering in 1882—just as Edison's Pearl Street station had become operational. Although only six students chose to enroll in the program that first year, it quickly gained in popularity. A year later, Cornell University became the second university in the United States to begin offering an electrical engineering program. These were the only two possibilities for "all things electrical" when Stone and Webster were making their college decisions.

Until 1902, MIT's course was part of the physics curriculum. It was conceived and led by Professor Charles Cross, head of the physics department. The physics connection was a natural one since electromagnetism studies usually were included in the study of physics. Cross was a strong advocate of electrical engineering, a dynamic educator, and well qualified. He had worked closely with Alexander Graham Bell, who used MIT's physics laboratory for experiments leading to the invention of the telephone. In 1884, Cross was a founder and vice president of the

[1] "The Electric Light Industry in America in 1887," *Telegraphic Journal and Electrical Review,* Vol. 20, February 11, 1887, p. 130.

American Institute of Electrical Engineers. It was at a meeting of the Institute in 1888 that Tesla unveiled his complete polyphase alternating current system.

The first year of MIT's curriculum was the same for all students; in the following two years, they concentrated on physics, mathematics, and mechanical engineering. It was only in their fourth and final year wthat "electricals" had a specific electrical engineering course of study. Also in their final year, each student completed a senior research thesis. The fourth-year course of study when Stone and Webster were students included such topics as Technical Applications of Electricity, Testing of Telegraph Lines and Dynamo Machines, and Electrical Testing and Construction of Instruments Laboratory (see Figure 6.1).

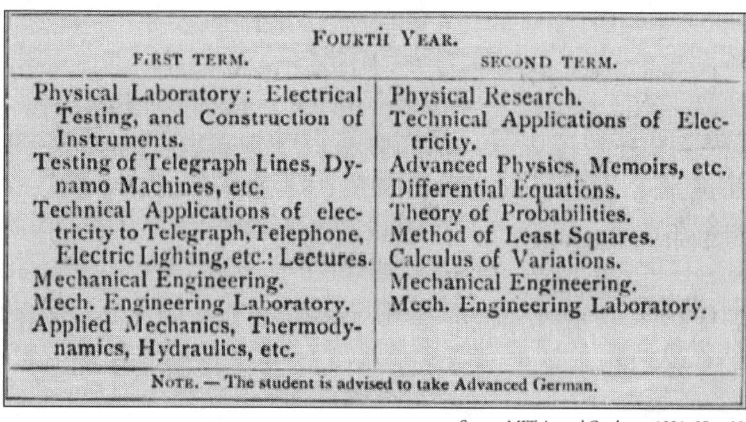

FOURTH YEAR.	
FIRST TERM.	SECOND TERM.
Physical Laboratory: Electrical Testing, and Construction of Instruments.	Physical Research.
Testing of Telegraph Lines, Dynamo Machines, etc.	Technical Applications of Electricity.
Technical Applications of electricity to Telegraph, Telephone, Electric Lighting, etc.: Lectures.	Advanced Physics, Memoirs, etc.
Mechanical Engineering.	Differential Equations.
Mech. Engineering Laboratory.	Theory of Probabilities.
Applied Mechanics, Thermodynamics, Hydraulics, etc.	Method of Least Squares.
	Calculus of Variations.
	Mechanical Engineering.
	Mech. Engineering Laboratory.

NOTE. — The student is advised to take Advanced German.

Source: MIT Annual Catalogue, 1884–85, p. 28.

Figure 6.1: MIT's Fourth-Year Course of Study in Electrical Engineering, 1884–85

Stone and Webster obtained Professor Cross's permission to work together on their senior thesis, titled *The Efficiency of Alternating Current Transformers*. Their 179-page thesis (half of the pages handwritten by each) was comprehensive and gave them a thorough understanding of AC systems that would serve them well in the future. In addition, the topic was timely and important. Westinghouse had redesigned the Gaulard-Gibbs AC transformer at the end of 1885 to make it manufacturable and efficient. He initially had one installed commercially in November 1886 in a Buffalo, New York, department store. These transformers were critical to the success of AC systems and were a factor in the AC-versus-DC battle of currents that raged in 1888.

DC system proponents asserted not only that AC current was dangerous but also that AC system transformers were inefficient. In his February 1888 booklet, *A WARNING FROM THE EDISON ELECTRIC LIGHT CO.*, published while Stone and Webster were working on their thesis, Edison stated that DC was preferable since "no translating, converting or other energy consuming apparatus intervene between the generating and the consuming units to waste a material percentage of the energy." [2]

According to Charlie Stone, Ted Webster and he frequently spoke while they were students about someday starting "some sort of organization for undertaking electrical business in this country."[3] Upon graduating in June 1888, they vowed to pursue this dream after obtaining some practical experience. Their entrepreneurial spirit and will to found their own business may have been at least partially influenced by their individual backgrounds. Both had fathers who were accomplished and self-made. Stone's father, Charles H. Stone, had modest beginnings as a gardener on a multimillionaire's estate, later becoming the owner of a successful wholesale produce business. He was the first Boston merchant to pack butter in ice for export to hot cli-

---

[2] *A WARNING FROM THE EDISON ELECTRIC LIGHT CO.*, February 1888, p. 23.

[3] David Keller, *Stone & Webster 1889–1989: A Century of Integrity and Service*, New York: Stone & Webster, 1989, p. 1.

mates. Webster was a member of a prominent Boston family. His father, Frank G. Webster, had risen from poverty to become a bank clerk, an employee of the Boston investment banking firm of Kidder, Peabody & Company from its formation in 1865, and a full partner of the firm while his son was completing his sophomore year at MIT. He would become the head of Kidder, Peabody in 1905.

The Kidder, Peabody association could have given Ted valuable insights into the electrical industry and its potential even before he became a student at MIT. Kidder was a major player in arranging financings for the railroad industry as well as for manufacturing businesses with large capital needs, such as electrical manufacturers and their related utility companies. Kidder had deep understanding of the electrical industry from Edison's earliest days. From about the time Ted was graduating from MIT until the 1920s, the investment banking industry was highly concentrated and dominated by an oligopoly of J.P. Morgan & Co.; Kuhn, Loeb & Co.; Brown Brothers; and Kidder, Peabody.

With his degree from MIT in hand, Ted Webster embarked on a European tour and then worked at Kidder, Peabody to better understand finance. Charlie Stone accepted a position as Elihu Thomson's assistant at the newly formed Thomson Electric Welding Company (Thomson later would become chief engineer at General Electric).[4]

---

[4] Professor Cross and Elihu Thomson knew each other well. Both were founders of the American Institute of Electrical Engineers in 1884. Thomson was a prolific inventor, amassing 696 patents during his lifetime. He invented electric resistance welding in 1886, and formed Thomson Electric Welding Company for its commercialization. Cross provided an endorsement of electric welding that was included in a brochure for the new company. Thomson and his colleague, Edwin Houston, also had

Matters coalesced about a year after graduation when the two young men met and, in Charlie Stone's words, "discussed a plan for the formation of an organization for electrical engineering and construction, and management of electrical properties. We even were courageous enough to feel that we might be able someday to have a banking department to buy and sell electrical securities."[5] A visit to their mentor, Professor Cross, followed.

At the time they had graduated from MIT, Cross had told them that he believed the time had arrived for when a bright young engineer might make a living on his own in the electric industry. Now, when reminded, he said, "Well, I perhaps did say that there might be enough work for one man, but I gravely doubt the possibility of two men starting in the business."[6]

Undeterred, Stone and Webster plowed ahead and, late in 1889, formed the Massachusetts Electrical Engineering Company with $2,600 (roughly $75,000 in today's dollars) borrowed from their parents. They renamed the business Stone & Webster three years later. It was one of the earliest electrical engineering consulting firms in the United States. They rented small offices in a downtown Boston building and set up shop, placing their desks back-to-back in an open area and signing all business documents and correspondence "Stone & Webster" rather than

---

formed the American Electric Company in 1880 to control patents for an arc-lighting system they had developed. Two years later, an investor group led by Charles Coffin assumed control of American Electric and renamed it the Thomson-Houston Electric Company. After Thomson-Houston was merged with Edison General Electric Company to become General Electric, Thomson was GE's chief innovator and was instrumental in establishing the famed General Electric Research Laboratory.

5 *Ibid.*, p. 11.

6 *Ibid.*, p. 10.

with their individual signatures. They continued these practices until their deaths.

The company's first big break came months later when, upon the strong recommendations of MIT President Francis A. Walker and Professor Cross, it secured its first major project: the design and construction of a direct-current system that would generate electricity at a dam in Maine and transmit it 1.1 miles to a paper mill. The firm subsequently completed a pioneering alternating-current transmission line development project for the same customer. More projects followed rapidly.

From the onset, Charlie Stone and Ted Webster maintained a close affiliation with MIT. They did not hesitate to make potential customers aware of this professional connection, which was not just with the president and faculty. They made a point of hiring MIT students for part-time assignments and MIT graduates for positions of major responsibility. They served on MIT boards and advisory committees and provided pro bono services. During 1911–16, they contributed extensive design services for and then constructed MIT's current campus on the banks of the Charles River in Cambridge.

Similarly, the young men did not hesitate to remind people of their family connection with Kidder, Peabody and that firm's financial strength. This connection resulted in the firm's next big break, in 1893, one that would alter and expand the scope of its work, change its character, and catapult it into a prominent position in the development of the electrical industry.

Until that time, electrical manufacturers, including Thomson-Houston and Edison General Electric (whose merger J.P. Morgan had engineered to form General Electric in April 1892) and Westinghouse, often had taken securities of newly founded utilities as partial payment for equipment and patent rights.

When another severe financial panic struck America in 1893, banks demanded payment on loans made to the manufacturers, who found it impossible to raise cash to meet those demands by marketing the utility securities that they held.

J.P. Morgan, to rescue General Electric and preserve his investment in certain other utilities, formed a syndicate to pay cash in exchange for those utility securities and then to manage or dispose of the stocks. Given Stone and Webster's credentials and referral and references from Kidder, Peabody and MIT, Morgan engaged the firm to appraise the acquired securities and arrange sales whenever possible. During the remainder of the 1890s and into the 1900s, Stone & Webster performed hundreds of appraisals throughout the United States and Canada. This gave the firm intimate, unique knowledge of the entire power industry. It also quickly opened a path toward managing and owning electrical properties.

In 1894, after detailed examination of the facilities and operations of a Nashville power and light company whose securities were held by Morgan's syndicate, Stone and Webster personally reported to Morgan that the company had excellent recovery possibilities. Morgan disagreed but said that if the young men felt strongly about their conclusion, they could buy the company for $60,000. Together, Stone and Webster only could muster $20,000 from their own funds. After being turned down by New York and Boston banks for the additional funds, they were able to obtain a $40,000 loan from Chicago financier J.D. Harvey. Having secured the funds, they took over management of the Nashville utility, repaid the loan within a year, and sold the company within five years for a total personal profit of $500,000 (over $14 million now). This alone made Stone and Webster wealthy men at an early age.

From engineering, construction supervision, and advising financiers on the valuation of utility securities they held, Stone & Webster expanded upon its Nashville experience and moved into reorganization and direct management of power utilities. Utilities realized that having Stone & Webster management brought stability and easier access to capital. By 1906, Stone & Webster was acknowledged as the leading service provider to the utilities industry.

The firm provided centralized management, engineering, and financial services to twenty-eight independent power, light, gas, and traction companies throughout the United States. It had a financial interest in each company, but each had its own officers, board of directors, and bank accounts. Stone & Webster in 1902 also had formed its own securities department, which placed securities of companies it managed.

By 1906, the firm was undertaking a number of major engineering projects. To handle the explosive growth and focus activities, its first subsidiary, Stone & Webster Engineering Corporation, was formed to manage all engineering, construction, and purchasing. By 1910, 14 percent of America's total electricity generating capacity had been designed, engineered, and built by Stone & Webster.

The company's reputation and influence in the electric industry continued to grow over the next two decades. In 1920, the company contractually provided management services to 59 utilities in 18 states. During that timeframe, nationwide power consumption was doubling every six years. By 1927, despite explosive overall industry growth and the emergence of new competitors, Stone & Webster had built 10 percent of the power plants operating in America and owned or managed 2.31 percent of the country's hydro and thermal-generating capacity.[7] And, at the same time, the firm was constructing Conowingo Dam on the Susquehanna River in Maryland. Conowingo was the second-largest hydroelectric project by power output in the United States after Niagara Falls and the largest powerhouse that ever had been built.

Just as utilities were drawn to Stone & Webster for management services, Stone & Webster's capabilities and reputation led the firm to become especially well known for designing and supervising construction of large hydroelectric projects such as Conowingo. Figure 6.2 highlights some hydroelectric projects undertaken by Stone & Webster over the years. It shows the evolution of the industry to facilities with larger and larger capacity and to dams along the nation's major rivers. Each project had a unique story. Some of those are featured in upcoming chapters.

Given the volume of Stone & Webster's work, one thing that set it apart through the 1910s and 1920s was the mobility of its construction teams. When construction-team supervisory personnel finished one project, they typically moved as a group to another project that was being started. Not only were the team members familiar with company practice, but they also knew each other and had worked together before. Some laborers also would move from project to project this way. This was a huge advantage in assuring high-quality, expeditious performance.

Construction men were tough, direct, independent people. Nonetheless, they developed loyalty to the company and to each other. Since projects typically lasted one to three years, this way of life, however, resulted in a nomadic existence for team members that, over time, was difficult for their families.

---

[7] Hughes, *Networks of Power*, p. 101.

Year	Project	kW	Comments
1904	Electron, Puyallup River, WA	28,000	First major West Coast project. 878-foot head, 10.4-mile flume.
1907	Taylors Falls, St. Croix River, MN	22,400	Supplied power to Minneapolis, 40 miles away.
1911	Hauserlake, Missouri River, MT	17,000	Replaced original failed steel dam.
1912	Goat Rock, Chattahoochee River, GA	30,000	60-mile power transmission. Original dam along stretch of river rapids.
1913	Big Creek, Big Creek, CA	147,000	Constructed in formidable remote terrain to supply electricity to Los Angeles. Originally had two powerhouses.
1913	Condit, White Salmon River, WA	14,700	Largest dam ever removed in the United States when decommissioned in 2011 to restore environment and salmon runs.
1913	Keokuk, Mississippi River, IA	135,000	Then the largest power plant and monolithic concrete dam in the world, and second longest dam in the world. Lock and dam enabled river traffic past rapids. Included a dry dock. See Chapter Seven.
1921	Caribou, North Fork of Feather River, CA	75,000	Part of extensive Feather River hydroelectric development. Power transmitted at 165,000 volts, highest in world then.
	Henry Ford Power Plants • St. Paul, Mississippi River, MN (1924) • Flat Rock, Huron River, MI (1923) • Green Island, Hudson River, NY (1923) • Iron Mountain, Menominee River, MI (1924)	 13,500 700 6,000 7,200	See Chapter Nine.
1925	Lower Baker, Baker River, WA	40,000	See Chapter Ten. Now 111 MW.
1926	Bartletts Ferry, Chattahoochee River, GA	65,000	Six miles upstream from Goat Rock Dam. These are two of six present-day generating stations along a 29-mile stretch of the river that together have 316 MW generating capacity.
1928	Conowingo, Susquehanna River, MD	252,000	See Chapter Eleven.
1931	Bagnell, Osage River, MO	176,200	Forms Lake of the Ozarks. Its construction provided jobs to 20,000 workers during the Great Depression.
1933	Rock Island, Columbia River, WA	78,300	First dam to span the Columbia River. Powerhouse expanded in 1953 to give capacity of 213,000 kW. Added another powerhouse in 1979 so that total power capacity now is 623,725 kW.
1955	Roanoke Rapids, Roanoke River, NC	104,000	Dam is 3,050 feet long, 72 feet tall.
1962	Rocky Reach, Columbia River, WA	728,000	Later expanded to 1,315,000 kW.
1963	Gaston, Roanoke River, NC	220,000	8 miles upstream from Roanoke Rapids and controls water flow to it.
1967	Cabin Creek, South Clear Creek, CO	324,000	Pumped storage facility. Difficult construction due to altitude and winter weather. Edison Electric Award winner.
1972	Northfield Mountain, Connecticut River, MA	1,168,000	Largest pumped storage facility in the world when became operational.

Figure 6.2: Some Stone & Webster Hydroelectric Projects

Stone & Webster rapidly became a large enterprise and expanded far beyond the electrical industry. During World War I, it constructed the mammoth Hog Island Shipyard near Philadelphia. Then the world's largest shipyard, it employed 35,000 workers. Stone & Webster built 122 vessels there in record time, using a new manufacturing model it had devised. Sections of the ships were fabricated at locations around the country, with final assembly at Hog Island. After the Great War, Stone & Webster provided engineering, design, and construction services for many types of industry, furnished supervisory services to public utilities and industrial companies, and offered general investment banking services. It constructed factories for Henry Ford, a sugar refinery, a rubber plant, pipelines, warehouses, manufacturing facilities, and airports. It built the 50-story RCA Victor Building (renamed the General Electric Building) in New York City and the 42-story Cathedral of Learning in Pittsburgh. It constructed what was, at the time of completion in 1924, the world's longest continuous tunnel, 18.2 miles through sandstone and shale rock, to double the Catskill water supply for New York City.

Nonetheless, the company never abandoned its roots in the power industry. In fact, a century after the firm's founding, a Stone & Webster researcher would note in 1989 that there had not been a single day since its first project, completed in 1890 in Maine, that the company had not been involved to some extent in hydroplant engineering and construction management.[8]

Even so, the work wasn't always easy or plentiful. The Great Depression severely impacted Stone & Webster. Construction essentially stopped in the entire United States. For several years,

---

8 Keller, *Stone & Webster*, p. 250.

completion of projects that had been initiated prior to the onset of the Depression in 1929 somewhat eased the decline in business. New jobs rarely came along. By 1932, construction activity was less than one third of what it had been in 1926. During 1930–33, employment at Stone & Webster's engineering and construction subsidiary dropped from 1,237 to a skeleton home-office crew of 263. Each person understood that his job would end upon completion of the project to which he had been assigned.

As the grip of the Depression tightened, the power industry was buffeted by lower electricity demand and no ability to finance projects. Franklin D. Roosevelt's election in 1932 ushered in his New Deal to create jobs via the establishment of the Tennessee Valley Authority to develop public power in the South, the Public Works Administration, and other agencies and programs. The Public Utility Holding Company Act of 1935 also led to massive restructuring and regulation of the electric utility industry. The dramatically expanded role of the federal government in these years fundamentally changed how hydropower would advance in the future.

After the dark days of the Depression, Stone & Webster recovered strongly as the company celebrated its fiftieth anniversary in 1939 and the United States prepared for and entered World War II in the early 1940s. In 1941, Stone & Webster achieved the highest volume of design, engineering, and construction work in its history.

Then, during the Second World War, the firm undertook a variety of large, critical engineering and construction projects. It is best known, however, for its involvement in the Manhattan Project, i.e., the race to develop the atomic bomb. Stone & Webster, among other things, designed and built the entire secret city of Oak Ridge, Tennessee, and constructed the electromagnetic separation plant there. This wartime involvement positioned Stone & Webster well for a leadership role later in the commercial nuclear power industry. It completed the nation's first full-scale, nuclear-powered electricity generation plant at Shippingport, Pennsylvania, in 1957. By the late 1970s, a significant portion of all nuclear energy in the United States was being generated at plants designed and built by Stone & Webster. The company also became a leader in petrochemical process engineering.

Nothing demonstrates the extent of Stone & Webster's involvement in the evolution of the electrical industry more than its presence in the State of Washington. In 1898, America had won just the three-month Spanish-American War, which turned the nation's eyes toward the Pacific region, resulted in the US acquisition of Guam and the Philippine Islands, and marked the debut of the country on the stage of world affairs.

It was then that Major Henry Lee Higginson of Lee, Higginson and Co., representing a group of small, troubled electrical utilities in Seattle, suggested that Stone & Webster look in to the situation there. Boston-based Lee, Higginson had been General Electric's principal investment banking firm dating back to its close relationship with Charles Coffin in the early Thomson-Houston days.[9] Major Higginson was an original member of General Electric's Board of Directors upon the company's formation. As previously noted, General Electric in the 1890s held stock of a number of distressed utilities that it had received in lieu of cash payments upon system implementation. The stocks

---

[9] Thomson-Houston's other principal investment banker had been Kidder, Peabody.

of the troubled Seattle utilities were among these. Higginson was forming a banking syndicate to acquire, manage, and consolidate these electrical properties.

After thinking about the matter, Edwin Webster said that with Washington's abundant natural resources, ideal conditions for hydropower, and railroad access from the East and with "the changes of trade conditions on account of the late war, I cannot help feeling that Seattle is bound to grow very rapidly."[10] His growth prediction would turn out to be accurate, as census data show (see Figure 6.3).

Census Year	Population
1890	42,837
1900	80,671
1910	237,194
1920	315,312
1930	365,583
1940	368,302
1950	467,591

Figure 6.3: Seattle Population Growth

Charles Stone traveled to Seattle late in 1898 and in 1899 to assess the situation first-hand and devise and implement a plan. In January 1900, the Seattle Electric Company was formed to acquire the various electrical-generation properties and scattered electric street railroads in the area and bring them under Stone & Webster management. Stone & Webster contracted with Seattle Electric to supervise management, development, and financing. It obtained an ownership position in Seattle Electric alongside financial houses and some local investors.

[10] *Ibid.*, p. 30.

Seattle Electric stock was used to acquire the various electrical-generation and street railroad properties. By 1907, sixteen Seattle electric railway and light and power companies were consolidated under Stone & Webster management. In addition, Stone & Webster assumed control and operation of other railway and electrical lighting and power utilities in surrounding western Washington areas. All Stone & Webster Washington holdings eventually were merged into two entities: Puget Sound Power & Light Company and North Coast Transportation Company.

After the Public Utility Holding Company Act of 1935 was passed, Stone & Webster elected to remain an engineering and construction services company and divested ownership of Engineers Public Service Company, its public utility holding company subsidiary, which controlled Puget Sound Power & Light and North Coast Transportation as well as several other large utilities (e.g., Virginia Electric & Power Company, or VEPCO). Stone & Webster nonetheless continued to build additional hydroelectric facilities in Washington.

From the time Stone & Webster became involved in Washington in 1900, it needed ever-increasing electric generation capacity to accomplish its goals. The first hydroelectric plant built in the state had been the Snoqualmie Falls facility, which became operational in 1899. The plant originally had six megawatts of installed capacity and transmitted its AC power 30 miles to Seattle. In 1903, Seattle Electric Company acquired Snoqualmie Falls Power Company and the power plant. Seattle Electric upgraded the plant in 1905 and added a second power plant in 1910. In 1904, Stone & Webster completed construction of the Electron facility on the Puyallup River. In 1905, the firm acquired the partially completed Nooksack Falls hydropower facility near Mt. Baker, intended to supply power to Bellingham, 90 miles north of Seattle. In 1911, another power plant was completed in the Seattle/Tacoma area on the White River. Upon completion of that project, Stone & Webster controlled all of western Washington's premier hydroelectric plants except Seattle's Cedar River municipal plant. In 1913, the Condit plant became operational on the White Salmon River. In 1925, the Lower Baker Dam was completed near Concrete. In 1933, Stone & Webster completed the Rock Island facility, the first dam to span the Columbia River.[11]

In addition to navigating a difficult business environment in the 1930s, Stone & Webster was experiencing another major shift. After forty-one years of being under the leadership of its two founders, it named another thirty-year company veteran, George Muhlfeld, chief executive officer at the end of 1930. Stone and Webster became chairman and vice chairman, respectively, and remained active in the company but with reduced attention to direct administration. Stone died in 1941 at the age of 74. Webster continued as chairman until retiring in 1946. He died in 1950 at the age of 82.

These leadership changes marked the passing of the entrepreneurial era for the electrical power industry and for the company that the founders had fashioned. At the 1989 centenary of the founding of Stone & Webster, its then-CEO, William F. Allen, Jr., observed that:

> Our founders and the team they put together were go-getters. They displayed great confidence in themselves and

[11] Stone & Webster also constructed the Upper Baker Dam and powerhouse in 1959. Major Rock Island upgrades became operational in 1953 and 1979.

> their company, and they never hesitated to move into uncharted waters when they saw opportunities. That kept them ahead of the competition and established a philosophy that is just as important today. There have been times in our history when we tended to stray from that course. Sometimes we had little choice in the matter, such as in the Great Depression of the 1930s when the entire nation slowed down. At other times, I think the deluge of business caused us to lose a little of the entrepreneurial spirit that keeps one looking always to the years ahead.[12]

Charles Stone and Edwin Webster were unique, fascinating individuals who were at the right place at the right time. They left an indelible imprint on a nascent industry that they helped build. They saw the light—and they helped light Americans' lives.

[12] Keller, *Stone & Webster,* pp. 343–344.

Fig. 6.4. Charles Stone (left) and Edwin Webster MIT Graduation Photo (1888)

Fig. 6.5. Stone (left) and Webster upon the Firm's 25th Anniversary (1914)

Fig. 6.6. Electron, Puyallup River Powerhouse with Flume in Background

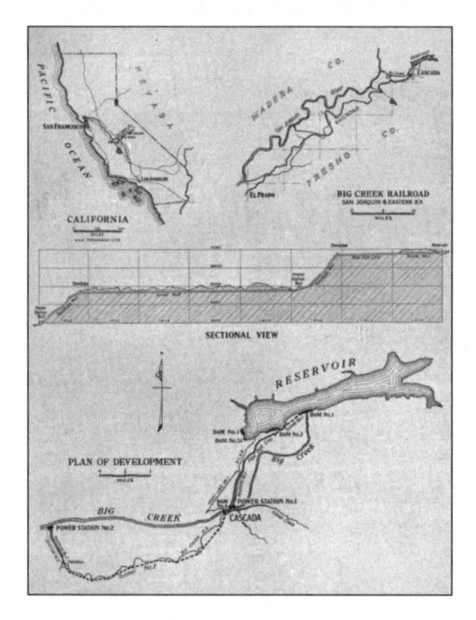

Fig. 6.7. Big Creek Development

Fig. 6.8. Condit Dam under Construction on White Salmon River

Fig. 6.9. Hauserlake Dam, Missouri River, Montana

Fig. 6.10. Rock Island Dam on the Columbia River

Fig. 6.11. Cabin Creek Pumped Storage Facility, Colorado

Chapter Seven

A Dream Come True

Dreams really can come true. One did for the residents of Keokuk, Iowa, in 1913.

Keokuk is located on the Mississippi River at its junction with the Des Moines River, about 175 miles upriver from St. Louis. The Des Moines River forms the border between Iowa and Missouri there. The town is named for Chief Keokuk, noted leader of the Native American Sauk tribe.

The Mississippi River is America's premier river, flowing 2,300 miles from its headwaters in Minnesota above Minneapolis to the Gulf of Mexico below New Orleans. It was the country's western border until Jefferson's Louisiana Purchase in 1803. It always has been a vital transportation artery. An 11-mile stretch of rapids above Keokuk became a serious obstacle as commercial river traffic grew in the 1800s. The rapids disrupted travel and limited access to lead mined at Galena, Illinois, and lumber from northern forests. Loaded steamboats often were delayed, frequently were damaged, and sometimes sank when transiting the shallow waters and strong currents. Soon the practice of lightering began: cargo was moved from steamboats onto small boats for passage through the rapids.

In 1837, the Army Corps of Engineers dispatched young lieutenant Robert E. Lee to chart and tame the rapids. Lee and his crew spent the next three summers attempting to cut a channel through the rapids. The project then was cancelled due to budget cutbacks. He pointed out that the river's immense waterpower should be harnessed. The river fell 22 feet through the rapids. The falling water's mechanical power could power saw, grist, or textile mills.

By the mid-1800s, Keokuk had become a bustling steamboat port of nearly ten thousand people. It served as a gateway for settlers moving westward through Iowa and was a staging point for Union troops during the Civil War.

After the Civil War, the Corps began constructing a canal paralleling and bypassing the rapids on the Iowa side of the river. The canal, completed in 1877, created a 5-foot channel and included three locks. A drydock also was built in Keokuk adjacent to the lock at the canal's lower end.

In 1893, leading citizens of Keokuk experienced the breathtaking electrical display at the Columbian Exposition in Chicago. The development of hydroelectric power at Niagara Falls starting two years later sparked their imagination: If Niagara Falls could be harnessed and spur economic development in the Buffalo area, why couldn't the mighty Mississippi River be harnessed to attract new manufacturing businesses to Keokuk? Why couldn't Keokuk grow and overtake St. Louis as a gateway to the West?

In April 1900, business leaders of Keokuk and its Illinois-side neighbor Hamilton formed the Keokuk and Hamilton Power

Company to make plans to dam the river and generate hydroelectric power. KHPC was enthusiastically supported by the residents of both Keokuk and Hamilton, and both towns appropriated public money to assist in KHPC's promotional efforts.[1]

KHPC engaged the prominent Chicago-based hydraulic engineer Lyman Cooley to confirm the possibilities for a dam.[2] The concept that emerged called for submerging the existing canal and locks and moving stretches of railroad tracks. Since navigation on the river had to be maintained, KHPC sought and eventually obtained Corps of Engineers endorsement of the proposed project. It would give the Corps a new lock and drydock at Keokuk, eliminate two upstream locks, and significantly improve navigation.

On February 9, 1905, President Theodore Roosevelt signed a law granting KHPC rights to construct and indefinitely maintain a dam across the Mississippi River at Keokuk.[3] The legislation required KHPC to construct, at its expense, a lock and drydock in conjunction with the dam. Ownership of the lock and drydock was to transfer to the Corps upon completion. The Act further required War Department approval of all construction plans and that construction commence within five years and be completed within ten years. KHPC had sent a delegation to Washington to lobby extensively for passage of the legislation.

---

[1] KHPC's organizers cobbled together $2,500 for its initial funding (about $80,000 in today's dollars), and Keokuk and Hamilton subsequently appropriated an additional $5,400. These funds were fully repaid by the dam's developer in 1908.

[2] He was best known for engineering the Chicago Sanitary and Ship Canal to reverse the flow of the Chicago River so that it would flow out of rather than into Lake Michigan.

[3] Fifty-eighth Congress, Session III, Chap. 566, 1905.

When news that Roosevelt had signed the bill reached Keokuk, fire bells rang and factory sirens blew as the town celebrated. KHPC said it was confident that dam construction could be completed within four years and that the region within 75 miles would soon become the most highly developed industrial area of the Midwest.[4]

Armed with project rights, KHPC needed a competent engineer or organization to undertake the project and the capital necessary to proceed. KHPC widely circulated a project proposal, or prospectus, to engineers and financial interests around the world. Only one expression of interest was received. It was from Hugh L. Cooper, who had supervised the just-completed construction of the Toronto Power Generating Station at Niagara Falls.

Cooper was a capable, self-taught civil engineer. Born in Minnesota in 1865, he began working as a laborer on bridge construction projects after graduating from high school. He was determined to become an engineer. As he later said, "I have had no college education. ... What I know has been gathered by night study and day practice."[5] By 1889, he was assistant chief engineer of the Chicago Bridge and Iron Company. He soon switched his focus to the nascent field of hydroelectricity and to designing and building hydroelectric plants in the United States and abroad. In 1905, he opened his own firm in New York City as the Toronto Power Niagara Falls project was winding down. He immediately was hired to design and build a major dam across the lower Susquehanna River at McCalls Ferry to supply pow-

---

[4] *Washington Times*, February 12, 1905, p. 9.

[5] Statement of Hugh L. Cooper, Hearings before the Senate Committee on Agriculture and Forestry on S. 3420, 67th Congress, Second Session, May 22, 1922, Washington, DC: US Government Printing Office, 1922, p. 707.

er to Baltimore. His involvement there ended In October 1907, when the project's development company filed for bankruptcy.[6]

A KHPC committee traveled to Niagara Falls to meet with Cooper in early September 1905. He quickly convinced them that he could manage the development of the Keokuk project and that the financial syndicate behind the Toronto Power Generating Station project would back him in doing so. KHPC eagerly granted him an exclusive option to arrange the financing.

Cooper was overly confident. The Toronto syndicate failed to support the Keokuk project. There was a major concern: No customers had been secured for the power Keokuk would generate, and the abundant cheap coal being mined nearby in the Mississippi Valley could make a power dam uneconomical.

It was imperative that customers be secured for the dam's hydropower. The best prospects were in the St. Louis area, which would require transmitting power nearly 150 miles from the dam site. This was farther than ever done before. Nonetheless, long-term contracts finally were signed in October 1908 to deliver 45 MW of electricity to St. Louis utilities, primarily to power electric street railways.

Fifty-eight financial groups turned Cooper down. The residents of Keokuk sank into deep gloom and despair. Failure appeared inevitable before Stone & Webster came to the rescue in 1909 as Cooper's option was on the brink of expiration. Stone & Webster formed the Mississippi River Power Company, with Edwin Webster as president and Hugh Cooper as vice president and chief engineer. The new company was assigned KHPC's project authorization. Due to Stone & Webster's expertise and

---

6 Construction subsequently resumed under new ownership. The dam, renamed Holtwood Dam, was completed in October 1910.

reputation, its endorsement of the venture enabled it to arrange $21 million in project financing ($580 million in today's dollars). This was the largest financial placement Stone & Webster had handled. The residents of Keokuk again were jubilant.

Cooper began some excavation work on January 10, 1910, less than thirty days before federal authorization for the project would expire. Major construction efforts did not commence until late in the fall.

The Keokuk project was a blockbuster. It was the first time the Mississippi River had been dammed below St. Paul, Minnesota. Harnessing the river's power was considered an engineering feat on par with taming Niagara Falls or the construction of the Panama Canal then underway.[7] When completed, the dam was longer than any other dam in the world except the Aswan Dam across the Nile River, and it was the longest monolithic concrete dam in the world. The powerplant was the world's largest single powerhouse and low-head hydroelectric facility.

The project works stretch 9,100 feet between bluffs on either side of the river and encompass the dam, powerhouse, lock, drydock, and an ice fender designed to protect the powerhouse and lock from ice floes. The monolithic concrete gravity main dam extends 4,649 feet across the river from the Illinois side. The run-of-the-river dam contains 119 spillways between piers on 36-foot centers. Waterflow over the gently curved spillways[8] is

---

[7] My paternal grandfather Paul H. Underwood, a long-time civil engineering professor at Cornell University, took a leave of absence from Cornell in 1911 to go to the Canal Zone and lead survey computing and mapping for the Isthmian Canal Commission.

[8] These technically are known as ogee-shaped spillways. The specific shape is chosen to smooth the flow of water overflowing the dam in high-water conditions with gates lifted.

controlled by the number of steel spillway gates that are open. The gates are operated by a gantry crane that moves on rails along the arched-span concrete service bridge atop the dam. The piers are about 55 feet tall and are anchored in the remarkably level and solid limestone river bedrock. The dam's backwater deepens the river for 60 miles upstream.

The powerhouse abuts the west end of the main dam and sits at a 110-degree angle to it. It is 894 feet long and is 177 feet tall from its foundation in bedrock below the river. Although Keokuk is a low-head hydropower facility with a normal head of 32 feet, the river's high flowrate of 100,000–200,000 cubic feet per second allows great power production. The powerhouse contains fifteen 25-cycle (or 25 Hz)[9] generators shaft-mounted to Francis turbines specially designed by Hugh Cooper. Total rated capacity is 135,000 kV.

In the early 1900s, electric power systems used many different frequencies. Higher frequencies worked better for lighting since they eliminated flickering; lower frequencies worked more efficiently with traction motors for streetcar systems. The United States had not yet settled on today's 60 Hz standard. The first generators at Niagara Falls were 25 Hz, so it is not surprising that 25 Hz was chosen for Keokuk. The Keokuk generators have been converted to 60 Hz since their initial installation.

The powerhouse originally was designed to be twice as long and contain fifteen more generator units. The substructure for

---

[9] An alternating current's frequency is the number of complete oscillations, or cycles, it completes in a second. The Hertz (abbreviated Hz) is the standard unit of measure for cycles per second . One Hertz is defined as one cycle per second. The Hertz is named after German physicist Heinrich Hertz, who conclusively proved the existence of electromagnetic waves.

the lengthened powerhouse was built, but the powerhouse superstructure extension was not added.

Never before had so much power been generated from single turbine runners. Rotating turbine-generator units need thrust bearings on their shaft. Traditional roller or ball bearings wore quickly and, even for smaller power units, typically required repair or replacement within months. The 550,000-pound weight of the rotating part of each turbine-generator unit made this a critical concern.

Engineering professor Albert Kingsbury had just invented a revolutionary thin-oil-film bearing that promised to be long-lived. The first Kingsbury bearing was installed for testing in June 1912 at the McCalls Ferry Dam, Cooper's previous project. The next Kingsbury bearing installations occurred a short time later at Keokuk. Some of those original bearings still are functioning flawlessly. Kingsbury bearings became a standard and made possible the design of much larger hydroelectric units such as those installed at Hoover Dam.

The new lock was adjacent to the Iowa-facing side of the powerhouse. Its 110-foot chamber width was the same as that of the locks being constructed in the Panama Canal. Its 400-foot length easily could accommodate any river traffic.[10] Its 40-foot lift was higher than any of the Panama Canal locks. The lock and the dam's reservoir obsoleted the canal and three locks that had been bypassing the Keokuk rapids and cut transit time by more than two hours. Beyond the lock to the Iowa shore was the new 150-foot-by-463-foot drydock. It was the largest fresh-water drydock in the world. Title to the lock and drydock was trans-

[10] Another lock constructed in the 1950s replaced this lock as river traffic increased and barge tow lengths increased. The new lock is 1,200 feet long. The 1913 lock and dry dock have been abandoned.

ferred gratis to the federal government when construction was completed.

The Keokuk project presented several significant construction challenges. Avoiding unnecessary delays was important. Cooper frequently pointed out that interest expense alone on the project's financing was more than three thousand dollars a day. Furthermore, because it was imperative that river traffic not be disrupted, Cooper built a trestle supporting a rail spur from the Iowa shore to the powerhouse construction area, with a steel drawbridge over the existing canal. Also, the river froze during the winter, with ice that was 2 to 3 feet thick. As the ice broke up in March and April, huge floes careened downstream.

Cooper and Stone & Webster divided construction responsibilities. Cooper was responsible for construction of the dam, powerhouse substructure, and navigation works. Stone & Webster designed and built the power station superstructure, electrical equipment, and all transmission lines. Cooper further segmented his work. Teams working from the Iowa shore were responsible for the powerhouse, lock, and drydock. Teams working from the Illinois side built the main dam itself.

Cooper drew upon his bridge-building experience for construction of the main dam. It was built section by section across the river. He designed a massive traveling cantilever crane to deposit cofferdam cribs, forms, and concrete as much as 125 feet beyond each just-finished pier and arch section. After completion of the arches and bridgeway, spillway bays were built one by one. Steel was used extensively for forms.

The last concrete in the dam was deposited on May 31, 1913. Less than a month later, the new lock was turned over to the Corps of Engineers. On July 1, Keokuk began sending electricity 144 miles to St. Louis over a 110,000-volt transmission line.

This then was the longest transmission line in the world,[11] and the receiving substation in St. Louis was the largest in existence.

A crowd of 35,000 people flocked to Keokuk to celebrate the dam's dedication in August. A mile-long parade wound down Main Street past cheering bystanders. Speaker after speaker spoke of a dream come true and praised those whose energy, faith, and extraordinary capacity had made the dam a reality. They foresaw Keokuk becoming a great manufacturing center in the days ahead. They extolled their hero, Hugh Cooper.

Stone & Webster managed the operations of the Keokuk Dam and associated power facilities until they were sold in 1925 to St. Louis's Union Electric Company (now Ameren Missouri). Today Keokuk plays a vital role in the reliability of Ameren Missouri's power grid by providing power needed quickly in a system emergency and helping meet peak-period demand.

Although commercially successful, the Keokuk hydroelectric facility never fulfilled the aspirations of the town's residents. The area never became a major manufacturing center. St. Louis and Chicago had too much of a head start and too many other advantages. Keokuk's population remains about the same as it was when the Keokuk project began.

Cooper went on to design and build a number of other major dams in the United States and internationally. He was made a colonel in the US Army during World War I and designed and supervised the construction of Wilson Dam across the Tennessee River. He later was retained by Soviet ruler Joseph Stalin to construct the Dneprostroi Dam across the Dnieper River in Ukraine. It was the largest hydroelectric plant in Europe.

---

[11] Keokuk briefly held this distinction. Stone & Webster's Big Creek project in California began transmitting power 248 miles to Los Angeles in mid-November 1913.

When Cooper died in 1937, *The New York Times* editorialized:

> He stood apart. With no formal technical education, he gravitated as naturally to engineering as artists to painting and sculpture. ... China, Egypt, Mexico, the United States, Soviet Russia, Chile—Colonel Cooper left his mark on them all in the form of engineering works which are as characteristic of our time as are the temples of the ancient world of theirs. And like the temples, his structures have the enduring quality that we associate with great masses. Long after this civilization has passed or merged into another, his magnificent dams will testify to the daring, imagination, and energy of an epoch dominated by the scientist and the engineer.[12]

The Keokuk hydroelectric project was unique for its scale and its innovations and for being the result of the vision and sheer willpower of the Keokuk community. This was hydropower's entrepreneurial time. The people of Keokuk and Hugh Cooper and the Keokuk Dam itself exemplified that spirit.

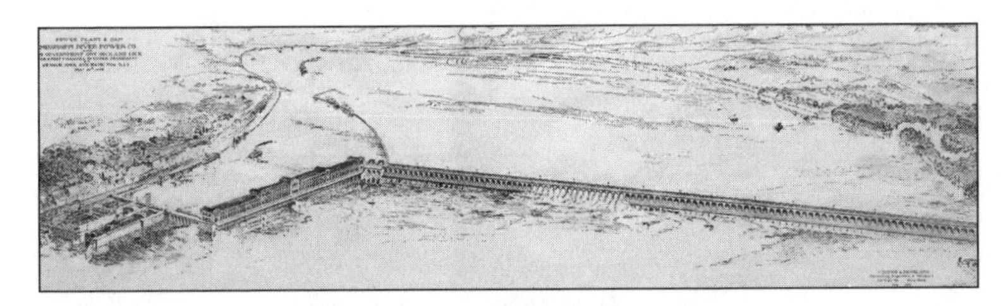

Fig. 7.1. Drawing of Keokuk Project Plan Prepared as Cooper Sought Financing in 1908. Some Details Changed Later.

[12] *The New York Times,* June 26, 1937, p. 16.

Fig. 7.2. Keokuk Dam and Powerhouse Panoramic View After Completion
Lock and Drydock in Lower Foreground

Fig. 7.3. Hugh L. Cooper in 1913

Fig. 7.4. Keokuk Main Dam Construction

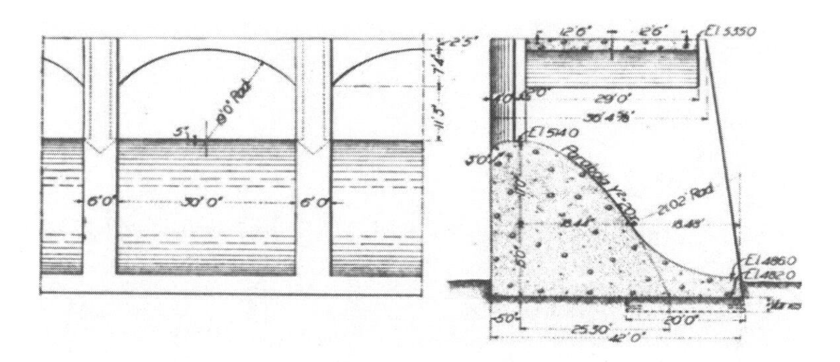

Fig. 7.5. Keokuk Main Dam Details Showing Spillway Shape

Fig. 7.6. Keokuk Powerhouse Construction

Fig. 7.7. Completed Keokuk Powerhouse

Fig. 7.8. Keokuk Generator Room

Fig. 7.9. Keokuk to St. Louis Power Transmission Line

Fig. 7.10. Two Steamboats in Keokuk Lock on Opening Day June 1913

Chapter Eight

Big Brother

The federal government played a minor role in hydropower's earliest days, but its role would evolve and expand greatly. By the 1930s and 1940s, Big Brother had become a major, even dominant, player. When the first Edison hydroelectric plant went into operation in Appleton, Wisconsin, in 1882, nobody thought about regulating this exciting, new, mere modification of an existing paper-mill mechanical waterpower system. The same was true for hydroelectric installations for years to follow. Eastern projects typically bolted onto existing waterpower systems and were highly localized. Early western projects usually were implemented near irrigation areas or water reservoirs. With the advent of AC power systems, such as the highly publicized system transmitting power 26 miles from Niagara Falls to Buffalo starting in 1896, and the move to larger and larger installations, hydroelectric projects started to attract regulatory attention.

The nation's waterways were its transportation arteries from the days of the early settlers and were critical to the westward expansion. They remain a key component of our infrastructure. The earliest federal role in water development and management dates to the 1820s, when the US Army Corps of Engineers was authorized to undertake work to improve the navigability of the country's waterways. This followed the 1824 *Gibbons v. Ogden* Supreme Court ruling, which found that Congress had the authority under the Commerce Clause of the Constitution to regulate navigation on interstate rivers. Among Corps activities were removing snags in major navigable rivers, deepening channels to maintain navigability, surveying rivers, and constructing levees for flood control.

The first federal legislation directly regulating hydropower was the Rivers and Harbors Appropriation Act of 1899, which made it illegal to construct any dam in a navigable waterway until the consent of Congress had been obtained and the plans had been approved by the Corps. These licenses were issued by special Congressional statutes following normal legislative processes. They were issued without a charge and with no time limit. Power companies initially had little difficulty obtaining permits, but that changed after Theodore Roosevelt suddenly became president of the United States in September 1901 following the assassination of William McKinley. He would remain president until 1909.

Roosevelt's Presidency captured the changing mood of the country. Whereas the last quarter of the nineteenth century had been dominated by big business's economic and political power, the Gilded Age of haves and have nots, monopolistic excesses, and the ravaging of precious natural resources, Roosevelt's Presidency ushered in the Progressive Era. Roosevelt earned a reputation as a trust buster, using regulatory reforms and antitrust prosecutions to achieve his goals.

An ardent outdoorsman, Teddy Roosevelt vigorously promoted the conservation movement and dramatically expanded the system of national parks and forests. In March 1903, he vetoed a Congressional bill that would have authorized a private group to construct a hydroelectric facility on the Tennessee River at Muscle Shoals (later the site of Wilson Dam). Congress was unable to override the veto, and the chair of the House Rivers and Harbors Committee pointed out that, in the legislation, the federal government would have given away a valuable privilege and "barter(ed) away for nothing that which is of greater value than the cost of the works themselves," and that it would have bestowed these privileges only upon the person who "comes here and passes a bill through Congress, thereby throwing the doors wide open for favoritism and monopoly."[1]

This was the first major restrictive move by the federal government to prevent special interests from usurping prospective waterpower sites. Several major influences would shape the ensuing debate over regulation of waterpower for years to come and end up stalling comprehensive water power legislation until 1920. These factors included:

- Conservationists believed that the country's natural resources belonged to the country's citizenry, that they were to be put to the best possible use, and that the benefits of their use should be shared with the citizenry. In the case of waterpower, conservationists believed that licensing for waterpower development should be available to all on equal terms via a fair and equitable licensing process, that the government should collect charges for the privilege of

1 *Congressional Record*, 57th Congress, 2nd Session, 1901–1903, Vol. 36, Part 3, p. 3072.

constructing hydroelectric projects in navigable waterways, and that there should be a time limit on any license granted. As Roosevelt said at one point, "These waterpower privileges are equivalent to many thousands of acres of the best coal lands for their production of power." He added, "To give away, without consideration, this, one of the greatest of our resources, would be an act of folly."[2]

- It was pointed out that 85 percent of the nation's waterpower resources were located on federal lands, upon navigable waters, or upon streams which cross or form international boundaries and that these lands and waters thus were under federal jurisdiction.
- Waterpower interests argued that water was free, that it was a reusable resource, that hydroelectric projects involved huge financial risk, that diminishing the potential financial rewards for risks assumed by developers was inappropriate, and that time limits on licenses made securing financing difficult if not impossible. They also saw federal fees as making hydroelectricity less attractive than coal-fired electricity generation in regions where coal was an option.
- Waterway interests jealously guarded navigability. The nation's waterways had been critical to the country's economic development and remained so. The rise of the country's railway system threatened waterway transportation. Waterway stakeholders envisioned continued improvement and expansion of navigable waterways to maintain their transportation viability versus rail. In the early 1900s, hydroelectricity of any scale was novel, and it affected local-

[2] *Congressional Record*, 60th Congress, 1st Session, 1907–1908, Vol. 42, Part 4, p. 3854, and *House Document No. 1350*, 60th Congress, 2nd Session, 1909, pp. 3–9.

ized areas. It was understandable that stream navigability then was of higher priority to many people than was waterpower.

- Utilities resisted federal regulation. By the early 1910s, largely as the result of Samuel Insull's impetus, state-level commissions regulated almost all electric utilities. Utilities understandably advocated the status quo.
- The economics of hydroelectricity and growing demand for electricity over time prompted the move to ever-larger waterpower projects. The potential impact of projects on their surroundings increased accordingly. The scope of larger projects required more established developers, invited consolidation of developers, and stoked fears of emerging monopoly as previously had happened with railroads.
- Presidential elections swayed the climate for regulation. Whereas Roosevelt was highly sympathetic to the conservationists, his successors were not as much so.

In June 1906, the General Dam Act of 1906 was enacted. Subsequent to passage of the Rivers and Harbors Appropriation Act of 1899, federal licenses for dam construction had been issued pursuant to special acts of Congress for each specific project. The 1906 Act was an attempt to delineate general terms and conditions that would apply to all such dam licensing bills (hence the name of the Act). The Act clearly was aimed at preventing the obstruction of navigable waterways by dams, and its provisions strongly discouraged waterpower development. Among its provisions were:

- requiring the approval of the Corps of Engineers and the War Department of the dam location and project plans and specifications;

- requiring permittees to construct, operate, and maintain locks and other navigation facilities without compensation at hydroelectric power sites whenever ordered by the federal government; and
- stipulating that the government could construct, operate, and maintain locks and other navigation facilities in the dam at any time and that lands needed for that purpose would be given without charge.

No time limit was placed on the duration of the grant, but construction of the project normally had to begin within one year and be completed within three years of the authorization.

The General Dam Act of 1910 amended the 1906 Act. It further impeded waterpower development by limiting grants of waterpower rights to a maximum of fifty years without provision for renewal. Furthermore, permits could be revoked at any time. The government could assess charges to reimburse its expenses for permit investigations or for restoring stream navigability. The Act made it virtually impossible to finance federally licensed waterpower projects.

The 1906 and 1910 Acts were practical failures in addressing the nation's needs for waterpower. Under these two Acts, between 1906 and 1920, a total of twenty-nine special Congressional acts passed and only eight dams were constructed, developing an inconsequential 100 megawatts of total power.[3] Other dams on rivers or streams not subject to federal jurisdiction, such as those that ran only within a particular state and/or were non-navigable, did continue to receive state licenses during this period.

---

[3] *Development of Water Power*, 66th Congress, 1st Session, Vol. 1, Report No. 180, Senate Committee on Commerce, September 12, 1919, p. 4.

The urgency for correcting the situation created by the 1906 and 1910 legislation became compelling as the nation's energy needs ballooned during World War I and as shortages developed in coal and oil supply due to labor strikes and other factors. There also was a growing realization that the existing statutes had left the nation's vast waterpower resources almost untouched, turning coal and fuel oil into the main sources of power. Steam power could be developed more quickly and easily with fewer legal restrictions and greater investment security.

In June 1920, the Federal Water Power Act was enacted. This pivotal legislation was intended to undo the waterpower development logjam by providing more effective coordination of the licensing of hydroelectric projects and by imposing license terms that would enable project developers to attract financing. The Act created the Federal Power Commission, composed of the Secretaries of War, Interior, and Agriculture, as the licensing authority for waterpower facilities. The Commission's work was to be performed by and through personnel of the departments of the three commissioners. The FPC was authorized to issue licenses with terms not to exceed fifty years for power developments on navigable streams, public lands, and forest reserves and at government dams. Upon expiration of a license, the government could take over a development by paying a licensee its net investment in the project. Alternatively, licensees could apply for renewal of licenses at the end of their term.

Licensees were required to pay annual fees for administrative costs of the Commission, rent for federal land, and assessments for increased value of waterpower sites by subsequent storage and flood control projects built upstream. Power companies also were to provide for navigation around dams where necessary.

The complexity of the licensing process and provisions in the

Act for government oversight of operations, maintenance, and safety incentivized hydroelectric project standardization during the 1920s and beyond.

Costs of obtaining and holding FPC licenses were high. Nonetheless, the 1920 Act, with its fifty-year licensing periods enabling project financing, along with an expanding postwar economy and accelerating demand for electricity, contributed to a waterpower building boom in the 1920s. Within the five years after passage of the Act, eighty-three projects licensed by the FPC, with a prospective installation of 1,975 megawatts of power, had been completed or were under construction. Ten of those, with a combined installation of 148 megawatts of power, had been started during fiscal 1925 alone.[4]

It should be noted that the Federal Power Commission labored under some real impediments. The three commissioners, each a cabinet secretary, had conflicting, overlapping and sometimes parallel authority over various phases of hydropower. There was only one actual employee, an executive secretary. All other personnel had to come from the staffs of the commissioners' departments. In addition, the 1920 Act stated that "the Commission may request the President of the United States to detail an officer from the United States Engineer Corps to serve the Commission as engineer officer." Internal bickering, divided organizational responsibilities, underfunding, staff shortages, investigative limitations, and lack of consistent policy hindered the operation of the Commission. In the 1920s, the commissioners met, on average, eleven times a year, with meetings each lasting about one half hour.

---

[4] *Fifth Annual Report of the Federal Power Commission,* fiscal year ending June 30, 1925, p. 2.

To accomplish its goals, the FPC relied extensively on and was influenced by the electrical industry it was regulating. A speaker at the 1921 annual convention of the Insull-dominated National Electric Light Association reported that:

> The Water Power Act of 1920, with the broad and liberal interpretation and construction placed upon its administration by the revised rules and regulations recently adopted by the FPC, and which were largely prepared by the Water Power Committee of the NELA, opens up the way for the speedy development of the water powers of the country.[5]

This lack of independence produced considerable criticism.

By the late 1920s, with the electrical industry becoming larger, consolidating via powerful trusts, and operating more and more across state lines, pressures mounted to strengthen the FPC and lessen its ties to the industry it regulated. In 1928, the FPC obtained its own independent staff after Congress authorized the transfer of borrowed departmental personnel to the FPC. In 1930, an amendment to the 1920 Act reorganized the Commission into an independent, five-member, bipartisan body appointed by the president with the advice and consent of the Senate.

The highly public downfall of Insull in the early 1930s and President Franklin D. Roosevelt's crusade to emasculate and dismantle power trusts led to passage of the Public Utility Holding Company Act of 1935. Supplementing PUHCA, Roosevelt pushed through Congress the Federal Power Act of 1935. This statute technically was an amendment and recodification of the

---

[5] *Proceedings of the 44th Convention of the NELA*, Vol 1, June 1921, p. 88. Samuel Insull's brother, Martin, was the Association's president at the time.

1920 Act. It extended the Commission's licensing authority to waterpower facilities on private land and non-navigable streams constructed by investor-owned utilities engaged in interstate commerce. It also expanded the powers of the Federal Power Commission to include regulation of all interstate electricity transmission and wholesale power sales. The Commission henceforth could set rates for interstate sales of electricity for resale.

Adapting to the times and ever-changing needs of American society over the years, additional amendments to the Federal Power Act continued to increase the role and licensing scope of the Commission (later renamed the Federal Energy Regulatory Commission) and to complicate the licensing and relicensing of hydroelectric facilities. Initially the licensing process focused primarily on project siting, dam design, and dam safety. Later, environmental concerns became paramount. The licensing and relicensing processes now require extensive planning and collaboration, including environmental studies, federal and state agency consultation, and public involvement. Licenses contain specific requirements about many aspects of hydroelectric projects, among them structures, equipment, water quality and other environmental issues, endangered species, wildlife and fisheries programs, operating procedures, stream flow, land management, recreational facilities, and historical and cultural issues. Displacement of people by new or expanded dam reservoirs also is a sensitive concern. Obtaining a license can take more than a decade of intensive effort.

The federal government not only has licensed and regulated water power projects since the earliest days of the electric industry but also has been building and operating hydropower facilities just as long. The US Army Corps of Engineers, the Bureau of Reclamation, and the Tennessee Valley Authority have been the primary federal players. Even though the Corps had been the federal custodian of the nation's waterways since the 1820s, it was the fledgling Bureau of Reclamation that built and operated the first government-owned hydroelectric facilities.

The Bureau manages water and power in the West. The region is largely arid, and, by the end of the nineteenth century, water was a major concern to support development of agriculture, settlement, and industry. Westerners clamored for the federal government to invest directly in irrigation projects, and pro-irrigation planks found their way into both Democratic and Republican platforms in the 1900 presidential elections.

Federally sponsored reclamation of arid western lands via irrigation gained a powerful proponent when Theodore Roosevelt became president in 1901. With Roosevelt's active support, the National Reclamation Act was enacted in 1902, less than a year after he took office. The Act formed the US Reclamation Service within the Department of the Interior and authorized "construction and maintenance of irrigation works for the storage, diversion and development of waters for the reclamation of arid and semiarid lands"[6] in sixteen (subsequently expanded to seventeen) western states. Construction was to be undertaken directly by the USRS (later renamed the Bureau of Reclamation, or BOR, and collectively called Reclamation).

Reclamation's first priority since its formation has been to deliver irrigation and drinking water, but hydroelectricity became its second major product. Off to a fast start, Reclamation in 1903 initiated its first five irrigation projects; twenty-three projects throughout the West had been authorized by the end of 1906. That year, Congress authorized Reclamation to take up

6 Act of June 17, 1902, 32 Stat. 388 (43 U.S.C. § 391), Sec. 1.

general development of hydroelectricity and to sell surplus power from water projects to towns and others and credit those sales revenues to the repayment of irrigation costs.[7]

Theodore Roosevelt Dam, built in Arizona as part of the Salt River Project, was far and away the largest of Reclamation's initial projects. When completed in 1911, the curved, 280-foot-high dam became the tallest stone masonry gravity dam in the world. It marked the end of the stone masonry dam era: concrete technology soon thereafter became the norm. Reclamation's first hydroelectric installation, a temporary 950 kW powerplant, went into service in Spring 1906 to support the dam's construction. In September 1909, a permanent powerhouse began delivering electricity to customers 65 miles away in the Phoenix area.[8]

The first Reclamation hydroelectric facility to deliver power commercially was the 900 kW Upper Spanish Fork powerhouse, which began operating in December 1908 as part of the Strawberry Valley Project in Utah. The power was used in drilling a 3.8-mile-long tunnel through mountains to connect a dam and reservoir in the Colorado River Basin with the Great Desert Basin below, in powering electric construction locomotives, and in powering irrigation pumps. Reclamation began selling the facility's surplus power to residents in a nearby community just months before power from Roosevelt Dam's powerhouse was delivered to customers in Phoenix.

These first powerplants initiated Reclamation's practice of selling electricity to defray costs of irrigation projects. It wasn't until the 1930s, however, that hydropower generation became

---

7 Town Sites and Power Development Act of 1906 (34 Stat. 116).

8 Rated capacity in 1909 was 3,000 kW. The capacity increased to 9,500 kW after the final generator was installed in 1916.

a principal Reclamation benefit due to the long, rancorous, national debate about whether private power production or public power production should be the rule. The mammoth Hoover, Grand Coulee, and Shasta dam powerplants that Reclamation built proceeded only after extensive Congressional debate. The phenomenal success of Hoover was a tipping point in favor of public power production.

Just as Reclamation's first hydroelectric facilities were beginning operations in Arizona and Utah, the Corps of Engineers under unique circumstances assumed ownership of its first hydroelectric powerplant at Sault Sainte Marie, Michigan. The St. Marys River is the only water connection between Lake Superior and Lake Huron. As it exits Lake Superior, the river drops 21 feet over rapids at Sault Sainte Marie. The border between the United States and Canada runs down the middle of the river. This waterway has been a critical passage since the country's earliest days, and was to French explorers and trappers even earlier.

The first lock was constructed at Sault Sainte Marie in 1797 to end the need for land portage of cargo. Other locks were added subsequently. By the late 1800s, the locks were heavily trafficked by freighters carrying iron and copper ore eastward from Lake Superior. (Today, the locks pass around ten thousand ships per year.)

In 1881, title to the locks was assumed by the US government, and the Corps of Engineers began maintaining and operating them. Private interests independently began developing hydroelectric facilities by the locks as early as 1885 to satisfy the needs of the adjacent Canadian and US cities of Sault Sainte Marie and of industry the developers hoped to attract. By the turn of the century, water canals had been or were being cut on both sides of the international boundary, waterpower plants were operating in Canada and on the river rapids on the American side of the border, and another expansive hydroelectric facility was being constructed along the riverbanks on the US side.

In 1905, the American and Canadian governments formed the International Waterways Commission. Among the commission's primary initial areas of focus were issues relating to the uses of water at Sault Sainte Marie. These included questions regarding water rights, water allocations, property ownership, and the potential impact of water diversions on Lake Superior water levels and vessel transit capability, all complicated by the location of an international border running along the middle of the St. Marys River. The commission recommended that their respective governments assume absolute control of all waters and lands necessary to improve and promote navigation in international streams.

Accepting that recommendation, Congress, in the 1909 River and Harbor Act, declared that federal ownership of all lands and property of any kind north of the St. Marys Falls Ship Canal and in the American St. Marys Falls rapids was necessary for navigation purposes and directed their condemnation. Because one of the two Soo power plants was located in the rapids area, the federal government via the Corps of Engineers suddenly found itself in the hydroelectricity business. It acquired that power plant from the Edison Sault Electric Company and allowed Edison Sault to continue to service the American city of Sault Sainte Marie by purchasing the excess power not needed to operate the locks. The facility's generation capability has been increased over the years to its current 17 MW. The 1909 Act also required the other, larger Soo power company, whose water canal and power plant were south of the river and locks and not in the rapids, to enter into a lease with the government for its use of canal water.

The next Corps of Engineers flirtation with hydroelectric facility ownership occurred in Minnesota at the navigational headwaters of the Mississippi River, where the Corps became embroiled in an intense rivalry between the cities of Minneapolis and St. Paul. Running downstream from Minneapolis, the Falls of St. Anthony drop the Mississippi River more than 100 feet before it arrives at downtown St. Paul. By the time of the Civil War, St. Paul, then the Mississippi's head of navigation, had become a busy port. Minneapolis understandably was eager to supplant St. Paul as the navigational end point.

After intense bickering and extended lobbying, Congress eventually authorized the Corps to build two lock-and-dam facilities that, together, would extend river navigation upstream to a point in Minneapolis below the St. Anthony Falls. Lock and Dam No. 2, the farther upriver of the two dams, was authorized in the 1894 River and Harbor Act; Lock and Dam No. 1 was approved in 1899. Neither of these low dams would support hydropower, which, at that point, was just beginning to become a concern. When Lock and Dam No. 2 became operational in 1907, Lock and Dam No. 1 was only approximately 20 percent completed.

Earlier that same year, Congress had approved creation of a 6-foot channel in the upper Mississippi River between St. Louis and Minneapolis. Locks and Dams No. 1 and 2 had been designed for a 5-foot channel and would need to be modified to comply. With hydropower quickly coming of age, pressure mounted to alter course and somehow build a high dam to exploit the area's hydroelectric potential while meeting the requirement for a 6-foot navigation channel to Minneapolis. In 1909, the Corps was authorized to study the hydropower question and suspended work on Lock and Dam No. 1, then 75 percent complete, pending results of the study.

The Corps-appointed study board concluded that only a high dam (approximately 30 feet) at the site of Lock and Dam No. 1 would make hydroelectric power economical. It further concluded that, according to Corps policy, the Corps alone could not build a high dam. It would have to confine itself to navigational considerations and partner with a private or municipal party to pay for the added high-dam costs.

The cities of Minneapolis and St. Paul proposed that they split the costs of the new high dam and share the resulting hydropower. In an unanticipated move that was contrary to policy, the Corps in January 1910 instead urged Congress to fund the entire project so that the Corps could control the electricity. Months later, the General Dam Act of 1910 was enacted. Two days thereafter, Congress approved the high dam for the Twin Cities and allowed the Corps to build the revised Lock and Dam No. 1 with a base for a hydropower plant. A nonfederal entity would have to develop the hydropower in the future. In a remarkable move, the top of Dam No. 2 was demolished—only five years after it had begun operation—to ensure navigational safety upstream of the new dam.

In 1923, Ford Motor Company was licensed to build a power plant atop Dam No. 1 to support an assembly plant it was constructing nearby. (We will learn more about this project in the next chapter.) In accordance with the Federal Water Power Act of 1920, Ford paid annual lease fees to the federal government. So, reflecting the battle about the federal role in hydropower that raged in the early 1900s, the Corps, in the end, enabled hydropower development on the Mississippi River at Minneapolis–St. Paul but did not own what it enabled.

World War I launched the Corps headlong into waterpower development. Nitrates were needed to manufacture high explosives. At the beginning of the war, Chile, with its natural deposits of sodium nitrate, or Chilean saltpeter, held a near monopoly on the world's supply of nitrates. Concerned that supplies could be cut off, Congress rushed to authorize construction of facilities for the fixation of atmospheric nitrogen. The enabling legislation specified that the facilities were to be constructed and operated solely by the federal government.

President Woodrow Wilson directed the Corps to construct two plants at Muscle Shoals, Alabama, to produce nitrate for wartime munitions and for fertilizer in peace time. The two plants employed different fixation processes—one called the Haber process, the other the cyanamid process. The Haber process was not successful, and the plant never was used. The cyanamid process required large amounts of electricity, so a hydropower dam (subsequently named Wilson Dam) was to be built on the Tennessee River nearby, as were coal-fired electricity generation facilities needed to support the nitrate plants until construction of the dam was completed.

Unfortunately, the first nitrates were produced at the cyanamid-process plant two weeks after the War ended in November 1918. At that time, construction of the dam was just beginning. So, with the War ended, the federal government, having spent a total of about $88 million (roughly $1.5 billion in today's dollars) on the project, had on its hands a white elephant with an estimated $8 million scrap value.[9] The nitrate plant was idled, and Newton Diehl Baker Jr., President Wilson's Secretary of War, immediately attempted to turn the plants over to private

[9] *The Cornell Civil Engineer*, Vol. XXX, No. 7, April 1922, pp. 95–96.

companies for fertilizer production. There were no takers. A bill introduced in Congress to create a government corporation to produce fertilizer at the plants failed.

As Warren G. Harding succeeded Wilson as president in March 1921, Congress defeated a $10 million appropriation for continuing construction of Wilson Dam. Harding's new Secretary of War, John Wingate Weeks[10], halted work on the dam on April 30 and promptly announced that, if any private corporations were willing to guarantee the government "a fair return" on its Muscle Shoals investment, he would recommend to Congress that the dam be completed and the entire project leased. At that time, the dam was about 40 percent completed, and the powerhouse was about 10 percent completed. Total expenditures and obligations on the dam had totaled $16.8 million to date. Secretary Weeks directed Major General Lansing H. Beach, Chief of the Corps, to solicit bidders for Muscle Shoals. As part of that effort, General Beach wrote directly to several leading industrialists, including Henry Ford.

Ford expressed interest in the project—the only interest in bidding on Muscle Shoals received by the government. General Beach became highly enthusiastic about the possibility of disposing of the Muscle Shoals properties to Ford. Ford, as we shall see in the next chapter, was a major proponent of hydroelectricity. He just had been issued a Federal Power Commission license (March 3, 1921) to build a power house at the end of a government dam across the Hudson River at Troy, New York, to provide power to a manufacturing plant to be constructed nearby.

---

[10] Weeks had made a fortune as founder of Hornblower & Weeks, a Boston investment banking and brokerage firm, before turning to politics. A US Naval Academy graduate, he served in the US Congress as both a Representative and a Senator before being appointed Secretary of War.

We already have learned of another power plant he would build at St. Paul, Minnesota.

What followed was a bruising battle between advocates of private ownership and public ownership of waterpower. The Muscle Shoals battle would continue unabated until the formation of the TVA in 1933.

Henry Ford made a surprise inspection tour of the Muscle Shoals facilities and the Wilson Dam site in June 1921. A month later, he submitted a bid for all the federal properties. It was a complex and controversial offer and was modified several times thereafter. The deal involved three basic components. First, Ford would pay $5 million (about $67 million in today's dollars) for the nitrate facilities, including the related steam power plants. He also would agree to manufacture a certain amount of fertilizer there annually, with a limit on his profits. Second, if the government were to complete Wilson Dam and another dam upstream that would regulate the amount of river water reaching Wilson, Ford would lease them for one hundred years, pay interest on the amount of capital required by the government to complete the dams, and make payments into a sinking fund to amortize the cost of the associated power equipment. Third, he would pay an annual property maintenance fee. The offer was nonspecific in a number of respects.

Partly to drum up and ensure enthusiastic local support for the offer, Ford, along with his friend and "technical advisor" Thomas Edison, again went to Muscle Shoals in December—to a hero's welcome. En route in his private rail car, the *Fair Lane,* Ford told reporters that his Muscle Shoals offer was motivated primarily by his desire to provide cheap fertilizer to American farmers. He also stated that he was contemplating using the excess power produced by the hydroelectric facility to manufacture aluminum, cloth, steel, and/or automobile parts. Upon his arrival in Alabama, Ford promised the large, eagerly awaiting crowd that he would make Muscle Shoals the "Detroit of the South" and subsequently announced he would employ a million workers and build a utopian, industrial city 75 miles long there. As he departed Muscle Shoals, he assured the send-off crowd that he was motivated by a desire to serve the public welfare rather than by an expectation of high profit.

No wonder he was wildly popular. The richest man in America was idolized by everyday Americans and viewed as someone who cared about them and could make their lives better (as exemplified by Ford's revolutionary $5-a-day minimum wage or the Model T bringing a new era to the common man). Rural Americans especially revered him. Impoverished Southern farmers suddenly saw him as their savior.

Playing upon his cheap fertilizer theme, Ford soon let it slip that Edison was working on fertilizer problems at his laboratories and already had informed him that they could give America a better fertilizer at a much better price than ever before.

Even though farming interests nationwide enthusiastically supported Ford's offer, opposition arose from many quarters as the proposal slowly worked its way through the War Department and then Congress. Among those opposing the offer were the power and chemical industries, Wall Street, and members of the Progressive movement, which had been evolving since the days of Teddy Roosevelt's administration. Each twist and turn was reported as hot news in newspapers across the country. Primary arguments raised against the offer included:

- The enabling legislation for the Muscle Shoals project (the National Defense Act of 1916) required that its nitrate plants be operated solely by the federal government.

- The offer violated a number of provisions of the Federal Water Power Act of 1920. For example, the Act limited licenses for waterpower development to a maximum of fifty years, yet Ford had proposed a term of one hundred years.
- It was not clear that Ford was obligated to manufacture nitrates regardless of whether the operations were profitable. Advances being made in the production of nitrates using the Haber process had the potential to render obsolete the cyanamid process of the Muscle Shoals plant.
- Many believed Ford would be getting too good a financial deal. Nebraska Senator Norris, the head of the Senate committee considering the bill, proclaimed that if Ford got Muscle Shoals, it would be "the greatest gift bestowed on mortal man since salvation was made free to the human race."[11]
- Ford could end up with a large amount of power in excess of what was needed to support fertilizer production. He could use this power however he wanted. It was argued that the government, in essence, would be subsidizing him, giving him an unfair competitive advantage if the power were used for his manufacturing businesses. Opponents asserted that power resulting from this government investment instead should be made available to the general public at fair and competitive rates.
- Ford's operation in isolation would not provide a multi-purpose regional solution.

War Secretary Weeks forwarded the Ford proposal to Congress in February 1922, along with his recommendation that

---

[11] Reynold M. Wik. *Henry Ford and Grass-roots America*, Ann Arbor, MI: The University of Michigan Press, 1972, p. 121.

the offer be rejected. He also recommended that construction of Wilson Dam be resumed regardless of whether the Ford proposal was accepted. General Beach of the Corps, however, simultaneously recommended that Ford's offer be accepted.

The offer was referred to the Military Committee in the House and the Committee on Agriculture and Forestry in the Senate. Opposition to Ford was led by the Alabama Power Company, both protecting its own interests and representing the entire power industry. Alabama Power submitted its own Muscle Shoals offer as House hearings began. A number of other proposals emerged over time. None prevailed. House consideration of the Ford proposal dragged on as various parties maneuvered until a bill authorizing acceptance eventually was approved in March 1924.

In the Senate, the Agriculture Committee was led by the wily Senator Norris, a staunch supporter of public power. Norris believed the power of Wilson Dam should be put to work for the public good, not for Henry Ford's private gain. Early on, his opposition to the Ford proposal made him hugely unpopular with—even despised by—Ford advocates. Norris received death threats from people who believed that Ford had been about to make them rich. Norris pressed on, becoming an expert on hydroelectric power matters, river navigation, flood control, the Southern economy, crop management, and fertilizer alternatives. He used his political skills to delay, delay, and further delay action on the Ford proposition while building opposition to it.

Nonetheless, with Ford's grass-roots support, through 1922 and into 1923, it appeared he would prevail. The richest man in the country also was a celebrity and arguably the most popular man in the country. Many Americans wanted him in the White

House. Ford-for-President clubs sprang up around the country in anticipation of the 1924 elections. A *Collier's* magazine national poll in July 1923 showed Ford substantially leading all candidates, including then-incumbent President Harding. The press and supporters in Washington asserted that, if Ford's Muscle Shoals bid were rejected, angry farmers would demand he be elected president.

When Calvin Coolidge became President upon Harding's death in August 1923, he undoubtedly was worried about his ability to win re-election a year later. On December 3, he and Ford met to discuss Wilson Dam. Ford emerged from the meeting saying he was "well pleased with his audience with the President."[12] Just over two weeks later, Ford announced his support for Coolidge in the upcoming elections.

This chain of events led many to conclude that a deal had been made by the two: Ford would not run for president if Coolidge delivered Muscle Shoals to him. About Ford's withdrawal from the presidential race, Norris said "it was considered by everybody in the country as a remarkable coincidence."[13]

Norris seized upon this situation to convene extended hearings in the Senate and to paint Ford as a big businessman in cahoots with a Wall -Street-controlled government. He further tried to give the impression of some kind of association with revelations about government corruption emanating from the simultaneously continuing Teapot Dome scandal hearings. In April 1924, he was able to produce a telegram from a Ford-associated reporter who had met privately with Coolidge in October relaying that Coolidge had said, "It is my hope that Mr. Ford will

---

12 *Ibid.*, p. 122.

13 *Ibid.*

not do or say anything that will make it difficult for me to deliver Muscle Shoals to him, which I am trying to do."[14] These revelations damaged Ford's image and impeded chances for approval of his Muscle Shoals proposal.

As Norris continued his stall tactics in the Senate, Ford withdrew his offer in October 1924, saying "a single affair of business which should have been decided by anyone within a week has become a complicated political affair."[15] Amazingly, the heated fight over the future of Muscle Shoals would drag on for over eight more years until the formation of the Tennessee Valley Authority in 1933.

Meanwhile, construction of Wilson Dam continued. Heeding War Secretary Week's February 1922 recommendation that dam construction be restarted, Congress authorized resumption effective October 1922. There was good reason to believe that the dam should be completed so that its hydroelectric potential could be exploited either by the government or by private parties in a region of the country in sore need of economic development. When completed, it would be a record-setting hydro facility. The 4,541-foot-long (about the length of Keokuk Dam), 137-foot-high dam would have twenty-one generating units. Its lock lift of 94 feet would be the highest then existing. And its neoclassical-style architecture, integrating themes of ancient Roman and Greek architecture into the modern structure, would be visually stunning.

Wilson Dam was designed and built under the supervision of Hugh L. Cooper. Cooper, together with Stone & Webster, previously had designed and supervised construction of the Ke-

---

[14] Preston J. Hubbard, *Origins of the TVA: The Muscle Shoals Controversy, 1920–1932,* Tuscaloosa, AL: University of Alabama Press, 2005, p. 127.

[15] *Ibid.*, p. 138.

okuk Dam on the Mississippi River (see Chapter Seven). The two dams had many similar design features.

The completed dam first generated power in September 1925, with fewer than half of its generators installed. With all generators in place, the facility eventually had an installed capacity of 436 MW. The total cost of building the dam was around $47 million (a walloping $700 million in today's dollars). Responsibility for the dam was transferred from the Corps to the Tennessee Valley Authority upon the TVA's formation in 1933. Power generated at Wilson Dam was sold to the Alabama Power Company while the dam remained under Corps responsibility.

The Muscle Shoals debate paved the way for the move to multiple-purpose regional hydroelectric facilities and the expansion of the role of the Corps of Engineers in government-owned hydropower. With passage of the Flood Control Act of 1928, the federal government assumed responsibility for managing the entire Mississippi River system. A program based only on levees was abandoned in favor of contained outlets and floodways, with a series of dams and reservoirs on every major tributary of the Mississippi. The Corps took over all engineering and construction. Via the Flood Control Act of 1936, Congress declared that flood control was a federal mandate and put the Corps squarely into reservoir construction. The Flood Control Act of 1944 allowed the Interior Department to sell power produced at Corps projects and authorized the mammoth, multipurpose Pick-Sloan Plan for the Missouri Basin.[16] Among the multipur-

---

[16] The Missouri River runs nearly 2,500 miles from its source in the mountains of western Montana before emptying into the Mississippi River north of St. Louis. The river basin drains about 530,000 square miles of land. Its lower reaches are subject to major flooding. At the direction of Congress, the Corps and Reclamation had separately submitted plans for

pose Corps dams on the Missouri River are Fort Peck, Garrison, Oahe, and Fort Randall dams.

Muscle Shoals was the precursor to formation of the third federal hydropower player—the Tennessee Valley Authority. (We will learn much more about the TVA in Chapter Fourteen.) The federal government now also operates four Power Marketing Administrations (e.g., the Bonneville Power Administration) within the Department of Energy to market surplus power from Corps and Reclamation facilities to their customers.

Today the Corps of Engineers, the Bureau of Reclamation, and the TVA together own nearly half of the country's installed hydroelectric capacity. According to 2014 data summarized in Figure 8.1, the 176 hydropower plants owned by these federal entities account for 49 percent of total hydropower capacity but only 8 percent of the plants. Publicly owned utilities, state agencies, and electric cooperatives own an additional 24 percent of capacity. The remaining quarter of the country's installed capacity, generated by 62 percent of the nation's plants, belongs to private owners.[17]

There's no doubt about it: The federal government rapidly became Big Brother in the hydropower market as the country's power needs burgeoned, the distribution of electricity went

---

development of the water resources of the entire basin. Lewis A. Pick of the Corps and William G. Sloan of Reclamation had authored the two plans. The plans subsequently were merged into the Pick-Sloan Plan. Initially, 121 projects were authorized, aiming to create 112 dams, 5.3 million acres of irrigation, 720 megawatts of hydroelectric generating capacity, and 1,500 miles of protective levees. President Franklin Roosevelt proposed absorbing the plan into a Missouri Valley Authority similar to the TVA. Attempts to create an MVA failed.

[17] *2014 Hydropower Market Report,* Washington, DC: US Department of Energy, April 2015, pp.12–13.

from localized to interstate networks, and the benefits of large, multi-purpose facilities became clear. With increasing federal regulation and federal ownership of waterpower facilities, the entrepreneurial earlier days of the industry were pushed aside. But we shall see that construction of larger and larger federal dams and power plants in the 1930s and beyond led to astounding engineering feats and human efforts.

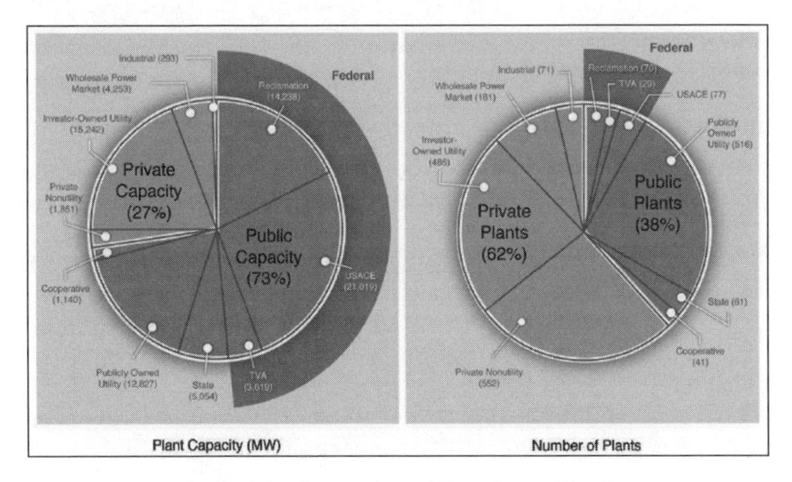

Figure 8.1: US Hydroelectric Plant Ownership Mix

Fig. 8.2. Reclamation's Roosevelt Dam under Construction circa 1910
Powerhouse at Base

Fig. 8.3. Powerhouse Generator Delivery in 1908, Strawberry Valley Project, Utah

Fig. 8.4. US Government Powerhouse, Sault Ste. Marie, Michigan
First Corps-of-Engineers-Owned Hydroelectric Installation

Fig. 8.5. Ford and Edison Inspecting Wilson Dam Site in December 1921

Fig. 8.6. Wilson Dam

Fig. 8.7. Senator George Norris circa 1920

Chapter Nine

It's a Ford

Henry Ford. What's the first thing that comes to mind? Model T automobile? Mass production? Famous industrialist?

Whatever your answer, odds are high that it wasn't hydroelectricity. Yet Ford was a strong advocate of hydropower and played a significant role in its evolution. He personally was responsible for thirty hydroelectric installations, large and small. He spotlighted national attention on hydropower's potential with his losing Muscle Shoals battle. Thomas Edison and he became the closest of friends.

Born in 1863 on a farm in what now is Dearborn, Michigan, Ford realized early that his interests were in things mechanical, not in farming. As a youth, he was intrigued by the workings of clocks and watches and became adept at repairing them. His fascination with waterpower began when he blocked a farm ditch to form a dam and constructed a makeshift water wheel connected to an old coffee mill. He and several schoolmates fed small pebbles into the mill to grind them into sand.

When he was sixteen, despite his father's objections, he left the family farm to become an apprentice machinist in nearby

Detroit. While working at the Dry Dock Engine Works, he supplemented his income by repairing watches for a jeweler. When he was nineteen, he returned to the family farm to help his father. He somehow managed also to work part-time setting up and servicing steam traction engines for Westinghouse. In 1891, he returned to Detroit to pursue his dream of developing a gasoline-powered horseless carriage. He supported himself while doing so by landing work as a night engineer for the Edison Illuminating Company of Detroit (later renamed Detroit Edison). He became chief engineer two years later.

Ford's work at Detroit Edison immersed him in the rapidly evolving early days of electrical utilities. He visited the Columbian Exposition in Chicago in 1893, where he absorbed the dazzling electrical displays. What impressed him the most, however, was a small gasoline engine used for pumping water. He took a mental photograph of it back to Detroit to give himself a clearer understanding of the direction of his own development efforts.

In 1896, Ford met Thomas Edison. It was a life-changing moment. It happened at the annual meeting of the Association of Edison Illuminating Companies. Forty-two men from eighteen Edison-affiliated electrical utilities around the country had gathered at the stately Oriental Hotel on Long Island to discuss topics of mutual interest. Ford had accompanied Alex Dow, president of Detroit Edison, to the meeting. After an afternoon session on the potential market opportunity for recharging vehicle storage batteries, Ford was seated for the meeting banquet at a table headed by the meeting's guest of honor, Edison himself. Dow and Ford were flanked by officers of the nation's other largest power companies, including Samuel Insull and the presidents of New York Edison and Boston Edison.

The topic of electric vehicles surfaced as the men talked after dinner. Dow pointed at Ford and announced, "There's a young fellow who has made a gas car." Ford had just finished hand building his first Quadricycle. This was the first vehicle he developed: a simple metal frame with a two-cylinder gasoline engine and four bicycle wheels mounted on it. Everyone wanted to hear more. Taking the chair next to the nearly deaf Edison, Ford began answering Edison's detailed questions and drawing explanatory sketches. When he finished, Edison brought his fist down on the table with a bang and said, "Young man, that's the thing; you have it. Keep at it." Ford later would say, "That bang on the table was worth worlds to me. No man up to then had given me any encouragement ... Here all at once and out of a clear sky the greatest inventive genius in the world had given me a complete approval."[1]

Edison sought out Ford the next day to ride back to New York together and to continue their discussions into a third day. Thus, the thirty-three-year-old Ford and the sixteen-year older Edison began an exceptionally close relationship (what we might call a bromance today) that would last thirty-five years until Edison's death in 1931. Near the end of Edison's life, Ford said of Edison, "I have come to know him, I think, rather intimately, and the more I have seen of him the greater he has appeared to me—both as a servant of humanity and as a man."[2]

Ford left Detroit Edison three years after meeting Thomas Edison to concentrate full-time on developing automobiles. His revolutionary Model T, the greatest automobile of all time, was introduced in October 1908. It put the world on wheels. By

---

[1] Henry Ford, *Edison as I Know Him*, New York: Cosmopolitan Book Corporation, 1930, p. 5.

[2] *Ibid.*, p. 16.

1918, about half of all American cars were Model Ts. An amazing 15 million Model Ts were built before production ended in 1927.

As demand for Model Ts soared, Ford began building a labyrinth of manufacturing facilities. He introduced assembly-line mass-production techniques in the Highland Park plant that opened in 1910 as the largest manufacturing facility in the world. Because it was far less expensive to transport auto parts and sub-assemblies on railroad cars than to ship finished automobiles, he began setting up branch assembly plants around the country. Cars were assembled at these branches for delivery to nearby markets. By 1914, there were fifteen branch assembly plants. In 1920, the mammoth vertically integrated Rouge factory complex began producing finished cars as well as millions of parts to be shipped to the company's network of branch assembly plants.

Locating factories on waterways and using hydroelectricity to power his factories were key strategies in Ford's expansion plans. In 1917, he ordered that all new plants be located on navigable waterways. Interruptions in the supply of materials, parts, or energy could be disastrous for his just-in-time manufacturing process. If railroads were too expensive, he could use barges. If coal miners went out on strike, he could use hydroelectricity instead of steam power.

Ford had an enduring interest in the use of hydroelectric power. In his mind, a flowing river or stream was free energy for the taking—wasted if it wasn't put to use. In 1910, Ford built his first small hydroelectric system on the Rouge River in Dearborn, Michigan. Three years later, on the adjoining 1,300 acres of farmland he had acquired just 2 miles from the farm on which he had grown up, construction began on Fair Lane, the Fords'

31,000-square-foot, 56-room mansion. With Edison's input, the original hydroelectric system was enlarged to power the house. Edison personally laid the cornerstone for the powerhouse in October 1914. The mansion included a suite of rooms reserved exclusively for Edison's visits.

By the time World War I ended, Ford had concluded that hydroelectricity was the most efficient, inexpensive, and least wasteful power source. A number of his manufacturing facilities soon utilized waterpower. Two of the first to do so added power plants to existing dams built by the federal government. These were a 13.5 MW power plant atop the Corps of Engineers' Dam No. 1 on the Mississippi River at St. Paul, Minnesota, to support a large assembly plant Ford was erecting nearby (see Chapter Eight for more about the dam) and a 6 MW power plant adjoining a dam built by the Corps across the Hudson River at Green Island, New York, for a parts production plant simultaneously being built there. Stone & Webster managed engineering and construction for these projects. Simultaneously, as highlighted in Chapter Eight, Ford also launched his unsuccessful effort to acquire the federal government's Muscle Shoals facilities and to lease Wilson Dam.

Upon the opening of the St. Paul assembly plant in 1924, Ford declared, "In providing for its gigantic power requirements, [Ford Motor Company] has seen the fundamental wisdom of harnessing the economical energy of waterpowers. Only in recent years has waterpower been extensively used in manufacturing, but, leading the way in this development, already ten Ford factories are hydroelectrically operated."[3]

---

[3] *Saturday Evening Post*, Advertisement, August 9, 1924, p. 55.

In 1920, Ford began opening small, hydroelectrically powered parts production plants along rivers and streams in rural southeastern Michigan. He labeled these plants "Village Industries." The plants allowed workers to maintain a farm while enjoying the pay and security of industrial work. Small, lightweight components manufactured in the plants were shipped to the huge Highland Park and Rouge production complexes nearby.

Many Village Industries plants were located at the site of old mills. Some utilized refurbished grist mill buildings. Before Ford's death in 1947, a total of nineteen Village Industries factories had been opened. Two had hydropower plants larger than 500 kW. Most were under 50 kW. Locating plants at old mill sites significantly reduced the installation cost of their hydroelectric systems.

Seventy-five potential hydropower locations interested Ford from 1918–45. Forty-five of those locations were in Michigan, but there also were sites in eleven other states.[4] Figure 9.1 recaps the thirty hydroelectric projects Ford went on to develop. A fascinating tale could be spun about each project, but the resulting book would be intolerably long. Instead, let's zero-in on the story of one facility: Iron Mountain.

Iron Mountain, Michigan, is 100 miles due north of Green Bay, Wisconsin, on the Menominee River. The river forms the border between Wisconsin and the Upper Peninsula in Michigan. The area is known for its vast timberlands, which was helpful for Ford: Model Ts were constructed largely of wood, for parts such as frame, floorboards, dashboard, steering wheel, and wheels. This translated into a huge need for hardwood.

---

[4] Ford R. Bryan, *Beyond the Model T: The Other Ventures of Henry Ford*, Detroit, MI: Wayne State University Press, 1990, p. 57.

Plant	Year	Comments
Major Facilities		
Green Island, Hudson River, NY (6,000 kW)	1923	Adjoined Corps dam. Produced radiators and springs. First license issued by the Federal Power Commission.
Hamilton, Great Miami River, OH (2,400 kW)	1920	Produced tractor transmissions, then automobile wheels.
Iron Mountain, Menominee River, MI (7,200 kW)	1924	Sawmill, kilns and wood distillation. Wooden auto parts.
St. Paul, Mississippi River, MN (13,500 kW)	1924	Atop Corps dam. Vehicle assembly and glass plant.
Personal		
Fair Lane Estate, Rouge River, MI	1915	Hydropower facility designed in collaboration with Edison.
Macon, Macon Creek, MI	1944	Completed but not opened due to Ford's failing health.
Village Industries		
Brooklyn, Raisin River, MI	1939	Horn buttons and starter switches. Employed about 130.
Cherry Hill, Rouge River, MI	1944	Lodging/employment for disabled WW II veterans. Radiator components, ignition and door locks and keys.
Clarkston, Clinton River, MI	1942	Seat covers and drill bushings.
Delhi/Huron Farms, Huron River, MI	1940	Property acquired; facility designed. Not completed.
Dexter, Mill Creek, MI	1933	Refurbished mill and rebuilt dam. Never completed.
Dundee, Raisin River, MI	1936	Copper welding points.
Flat Rock, Huron River, MI (700 kW)	1923	Head and tail lamps. Peak employment 1,200.
Goose Creek, Goose Creek, MI	1941	Under construction but not completed.
Hudson Mills, Huron River, MI	1942	Under construction but not completed.
Manchester, Raisin River, MI	1940	Dashboard cluster assemblies/ammeters. 150 employees.
Milan, Saline River, MI	1938	Ignition coils and soybean processing. 144 employees.
Milford, Huron River, MI	1939	350 employees. Carburetor production.
Milford-Pettibone Creek, MI	1938	Second powerplant supporting carburetor factory.
Nankin Mills, Rouge River, MI	1920	Manufactured screws. Employed 24.
Newburg, Rouge River, MI	1935	Produced the twist drills used in all Ford plants.
Northville, Rouge River, MI	1921	Valve manufacturing facility. 300 employees.
Phoenix, Rouge River, MI	1922	Employed only single women. Made generator parts.
Plymouth, Rouge River, MI	1923	Employed 23. Taps and other manufacturing tools.
Saline, Saline River, MI	1938	Soybean oil for paints. Parts for WW II aircraft engines.
Sharon Mill, Raisin River, MI	1938	Smallest plant. 14 employees. Produced cigar lighters.
Tecumseh/Hayden Mills, Raisin River, MI	1935	Soybean processing for paints and plastics. Employed 25.
Waterford, Rouge River, MI	1925	Up to 210 people making precision inspection gauges.
Willow Run, Huron River, MI	1939	Made locks and keys.
Ypsilanti, Huron River, MI (2,100 kW)	1932	738 employees. Produced generators and starters.
Sought but Not Developed		
Muscle Shoals, Tennessee River, AL	1925	Wilson Dam completion doubtful without Ford offer.

Figure 9.1: Ford Hydroelectric Facilities

At the end of World War I, lumber from northern woodlands was shipped to Highland Park, made into parts, and then shipped out to branch assembly plants. Ford believed that lumber dealers were taking advantage of his need for high-quality hardwood by price gouging. He also wanted to stop paying long-distance freight charges on the 40 percent water weight of green wood and on the scrap wood that was left over after cutting parts from boards. He decided to establish a sawmill and autobody parts factory near the sources of hardwood and ship the parts directly to his assembly plants.

Henry Ford's first cousin, Minnie Flaherty, with whom he was close, was married to E.G. Kingsford. The Kingsfords resided in Iron Mountain, where Kingsford was a land and timber appraiser, broker, and developer, and a local Ford dealer. It was he whom Ford asked in 1919 to quietly acquire large timberland tracts on his behalf. Kingsford did this very successfully. Within a year he had acquired 313,000 acres (later to grow to over 400,000 acres, or more than 625 square miles). By this time, over a million board feet of lumber were required daily for Model T production.

In July 1920, Ford began construction of the Iron Mountain plant.[5] It included a mammoth saw mill, drying kilns, an automobile body plant, and a wood distillation facility. Kingsford oversaw facility construction and subsequently managed the operation. At the time, it was the largest operation of its kind in the world. It still is the largest industrial enterprise ever undertaken on the Upper Peninsula of Michigan. By 1925, the facility employed nearly 8,000 people, was making 69 different auto-

[5] The plant actually was constructed on a site just south of Iron Mountain that later was incorporated into what now is the City of Kingsford.

body parts, and was producing about 350,000 wooden parts a day. In 1929, the facility added production of the famous Ford Woody station wagon body. During World War II, more than four thousand wooden CG-4A invasion gliders were built there for the US Army.

Power for the plant at first was generated by huge steam boilers burning oil, wood, and refuse. Given Ford's penchant for waterpower and decisions to expand the size of the Iron Mountain plant, it was decided early on to dam the nearby Menominee River and construct a power station to generate hydroelectricity to supply power to the plant. A construction contract was awarded to Stone & Webster. It was one of the four hydropower plants built for Henry Ford by Stone & Webster, the other three being St. Paul, Green Island, and Flat Rock. Stone & Webster's construction superintendent was my grandfather.

Just thirty-nine years old at the time, George Jessup was a gruff, direct, highly practical project manager. He had electrical power facility construction in his blood. While an engineering student at Cornell University, he had been a foreman on the construction of the university's original hydroelectric powerhouse. After Iron Mountain, he would go on to manage the construction of ten other major US waterpower projects.

Dam and powerhouse construction began at Iron Mountain in May 1923. With intense pressure to complete the project as quickly as possible, despite highly adverse winter weather conditions, the facility began supplying power thirteen months later. The dam across the Menominee River was 849 feet long overall, created a hydraulic head of 30 feet, and backed Cowboy Lake up the river 9 miles to another dam at Twin Falls. The poured-concrete center section contained 10 steel radial gates, each 20 feet wide and 14 feet high. The interior walls of the 119-foot-long, concrete powerhouse were a soft tan color with green trim; the flooring was inlaid with red tiles. Francis vertical turbines connected to three Allis-Chalmers generators produced an installed capacity of 7.2 MW of electricity. Electricity was fed to the Ford plant through a 2-mile system of underground conduits. When Ford closed its Iron Mountain operations in 1951, this hydroelectric facility was sold to Wisconsin Michigan Power Company (now WEC Energy Group). Its operation continues today.

In August 1923, while construction on the Iron Mountain dam was proceeding furiously, Henry Ford and his close friends Thomas Edison and Harvey Firestone, who dubbed themselves the Vagabonds,[6] arrived in Iron Mountain on one in their series of auto-camping summer road trips.[7] Between 1915 and 1924, the Vagabonds and their entourage trekked to various locations across the country. Newspapers ran features on their latest adventures, and newsreels of their exploits were shown in movie theaters nationwide. They would caravan by automobile during the day and set up camps at night to rough it outdoors.

For the trip to Iron Mountain, the Vagabonds boarded Ford's 200-foot-long personal yacht to Escanaba, Michigan, and then caravanned in chauffeur-driven Lincolns along with their supply vehicles across the Upper Peninsula. At Iron Mountain, they

---

[6] The Vagabonds group also included the famous naturalist John Burroughs (1837–1921).

[7] Auto camping, promoted by the adventures of Ford and the Vagabonds, became popular in the early 1900s with the advent of the automobile: "The motor vehicle quickly became integral to the camping experience, not only for what it could transport but also as an extension of the campsite." Martin Hogue, "A Short History of the Campsite," *Places Journal*, May 2011. https://placesjournal.org/article/a-short-history-of-the-campsite/

were hosted and joined by E.G. Kingsford. While there, Ford and Edison went to the dam construction site to view progress. Jessup vividly recalled the occasion and, when I was in high school, told me that "Edison was silent, but Ford immediately started telling me what to do. He didn't know a damned thing about how to build a dam. I just kept nodding my head and saying, 'yes, Mr. Ford.' Afterwards, I told my men to keep right on doing what they had been doing."

Ford detested waste of any kind. The Iron Mountain sawmill and wooden parts production operations generated a lot of waste. Three months after the Iron Mountain hydroelectric complex was finished, a distillation facility that converted the waste wood into wood alcohol, wood tar, gas, oil, and charcoal was opened. Each ton of wood waste produced 610 pounds of charcoal. Ford had had an idea: The charcoal would be crushed, mixed with starch, and compressed into useable briquets. Edison worked with Ford to design a briquet facility that was placed next to the distillation building. So began Ford's charcoal briquet business. Branded barbecue accessories and packages of the popular briquets were sold in Ford dealerships across the country. The business grew. A group of investors bought Ford Charcoal in 1951, and the name of the company was changed to Kingsford Products in honor of E.G. Kingsford and the town next to the factory. Kingsford Charcoal now is a subsidiary of The Clorox Company.

It is difficult to overestimate Henry Ford's vision, his intellectual curiosity, his drive, his skill in making big things happen, or his influence in shaping both modern industry and the evolution of hydropower. His story also shows how interconnected the world of hydroelectricity movers and shakers was from the early 1900s up until World War II, as evidenced in part by the fascinating interactions of Edison, Insull, Stone & Webster, Ford, Senator Norris, the Federal Power Commission, the Corps of Engineers, the Congress, and a string of US presidents.

Fig. 9.2. Ford with Original Quadricycle and 10 Millionth Model T in 1924

Fig. 9.3. Ford Fair Lane Powerhouse Cornerstone
Laid by Thomas Edison in 1914

Fig. 9.4. St. Paul Mississippi River Dam, Ford Powerhouse and Assembly Plant

Fig. 9.5. Green Island Lock, Dam and Power Plant

Fig. 9.6. Stone & Webster Ford Waterpower Projects Advertisement

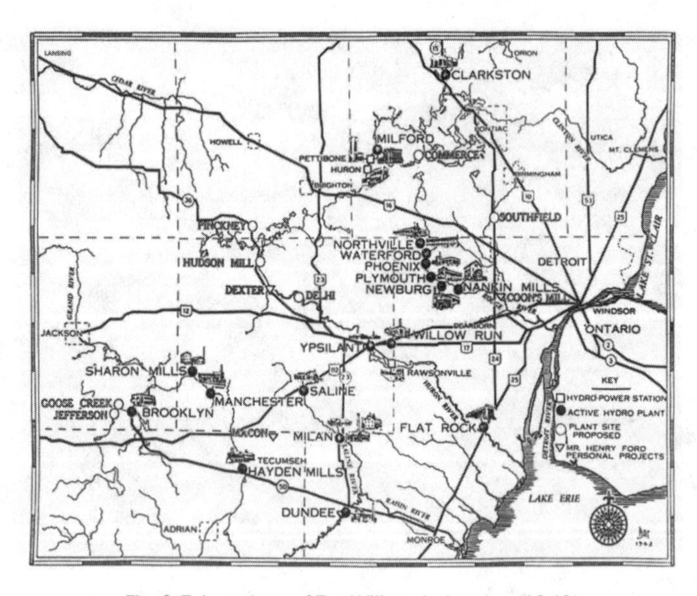

Fig. 9.7. Locations of Ford Village Industries (1942)

Fig. 9.8. Ford Village Industries Advertisement

Fig. 9.9. Ford Dam at Iron Mountain, Michigan

Fig. 9.10. Ford Plant at Iron Mountain/Kingsford

Fig. 9.11. The Vagabonds Pose atop a Waterwheel (1918)
Thomas Edison, John Burroughs, Henry Ford, Harvey Firestone (L to R)

Fig. 9.12. Vagabonds Travel to Iron Mountain in 1923 Aboard Ford's Yacht Sialia *Then by Automobile Caravan*

Fig. 9.13. Vagabonds Firestone, Ford, and Edison plus E.G. Kingsford (L to R) at Iron Mountain, August 1923

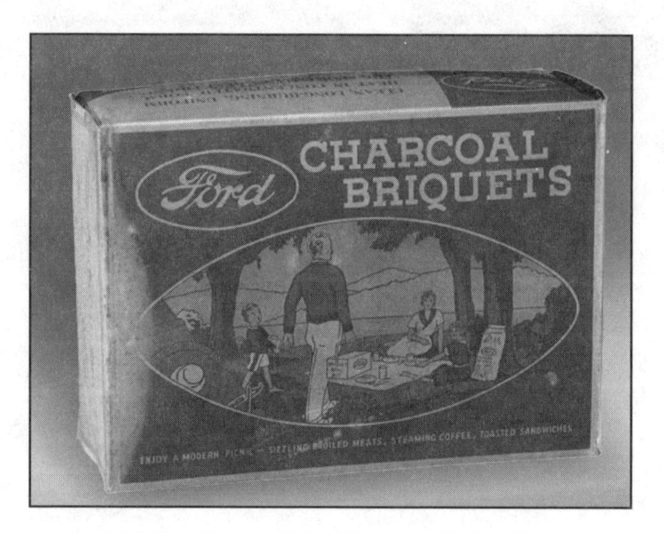

Fig. 9.14. Ford Charcoal Briquets Produced at Iron Mountain Later Renamed Kingsford Charcoal

Chapter Ten

Power Plays

While Ford's offer for Muscle Shoals was being debated in Washington, DC, big things were happening hydroelectrically out west in the other Washington: the State of Washington. From the earliest days of electricity onward, Washington State was hydropower nirvana. It is a geologically rugged state. Two mountain ranges, the Cascades and the Olympics, dominate the topography. The plentiful mountains are high, and water flowing from their snow-packed peaks is abundant. Studies in the 1920s concluded that Washington had the greatest hydroelectric power potential of any state. Today, Washington generates over 25 percent of the nation's total hydropower, with hydropower fulfilling 70 percent of the state's energy needs.

Demand for electricity in western Washington grew rapidly during and after World War I. Existing facilities could not meet the demand, and utilities decided to build new hydroelectric projects. By then, the most obvious, inexpensive potential waterpower sites near urban areas where loads were concentrated already had been exploited. Thus began the move to develop larger-scale, costlier remote sites. As shown in Figure 10.1, three significant projects were completed between 1924 and 1926: one by Seattle's municipal utility, one by the Tacoma municipal power utility, and one by the larger and more geographically extended Puget Sound Power and Light. Investor-owned Puget Sound Power and Light, as we learned in Chapter Six, was controlled by Stone & Webster.

Owner	Project	Project Description	MW	Completed
Seattle City Light (municipal)	Gorge Dam, Upper Skagit River	30-ft diversion dam, 11,000-ft tunnel, 294-ft head	60	1924
Puget Sound Power and Light (investor owned)	Lower Baker Dam, Baker River	293-ft-tall arch dam, 1,600-ft power tunnel	40	1925
Tacoma Power (municipal)	Cushman Dam No. 1, North Fork of Skokomish River	275-ft-tall concrete arch dam	43	1926

Figure 10.1: Significant Washington Hydroelectric Projects, 1924–26

These projects highlighted the fierce, ongoing battle being fought between municipal utilities and investor-owned utilities in western Washington. In 1900, Stone & Webster had taken steps to consolidate all of Seattle's then-existing streetcar and electric utility firms into the Seattle Electric Company. Not everyone enthusiastically greeted this move. Belief that the Eastern money behind Seattle Electric would siphon dollars away from Seattle to the detriment of its residents, Seattle Electric's high rates, persistent service problems, and reports of underhanded political dealings spawned a movement to establish a municipal power company. A 1902 Seattle Chamber of Commerce resolution supporting municipal power included the following statement:

> Under the existing conditions, Seattle is at the mercy of one company which does not hesitate to take advantage of its monopoly.... An exorbitant charge is made to factories ...

> a charge far in excess of what is taxed in Tacoma or Everett for a similar service. This condition of affairs could not last 24 hours had the city a municipal lighting plant ready to furnish power at a minimum of cost.[1]

In 1902 and 1903 elections, voters approved the creation and financing of a Seattle municipal power company, Seattle City Light, and the construction of a municipal hydroelectric plant on the Cedar River. By 1910, the municipal system was supplying about one third of the electricity consumed in the city.

In 1912, Seattle Electric was consolidated into Stone & Webster's newly-formed Puget Sound Traction, Light, and Power Company (re-named Puget Sound Power and Light Company eight years later). The war between Seattle City Light and Puget Sound Power and Light over Seattle's power grew intense as each maneuvered for position in western Washington's most concentrated center of demand.

In 1911, the State of Washington created a Public Service Commission to regulate investor-owned utilities. Municipal utilities, however, were not subjected to state regulation. This led Stone & Webster to complain that "the municipal plant in Seattle has had perfect freedom to conduct its business as it saw fit. Discrimination in rates, inducements to obtain business, intimidations in the matter of building permits and questionable accounting methods are abuses that this freedom from regulation has allowed."[2] In 1915, after extensive lobbying by Stone & Webster, legislation was passed prohibiting municipally-owned

---

1 Charles David Jacobson, *Ties That Bind*, Pittsburgh, PA: University of Pittsburgh Press, 2000, p. 114.

2 *Chronological History of the Puget Sound Power and Light Company and Predecessor Companies, 1885–1938*, company pamphlet, 1939, p. 7.

utilities from extending their distribution networks outside of a city's boundaries.

Seattle City Light's driving force was the legendary J.D. Ross, a self-taught engineer who had overseen construction of the Cedar River power plant. In 1911, he was named superintendent of City Light and held that position almost continuously until his death twenty-eight years later. Foreseeing the increase in demand for electricity in store for Seattle and recognizing that waterpower was the logical source of electricity to meet that demand, Ross first attempted a major upgrade of the Cedar River facility. A new, 215-foot-high masonry dam designed to increase generating capacity by 2.5 times was completed in 1914. It was built on glacial moraine after cautionary initial engineering studies were ignored; consequently, it was plagued by seepage and was practically useless as a hydropower reservoir. The embarrassed Ross scrambled to investigate every other possible water site within 150 miles of Seattle.

Stone & Webster, meanwhile, attempted to prevent City Light from ever getting another hydro site. The company hastily bought two separate sites while City Light was negotiating for them. Stone & Webster also had tied up rights to another promising site 100 miles northeast of Seattle on the Skagit River, having obtained a federal permit for the site without developing it. It appeared that City Light was boxed in.

But Ross soon struck back.

The Skagit River flows from British Columbia to the Puget Sound, draining the North Cascades. As the upper portion of the river flows through the Washington National Forest, it descends 700 feet while passing 15 miles through a narrow gorge of solid granite. This remote, rugged, nearly inaccessible location had long been considered to have the greatest hydropower potential on the west slopes of the mountains. The US Department of Agriculture held authority over power sites in National Forests. Although Stone & Webster had secured rights to develop the Skagit River for power, the company's permit included a requirement that construction begin within a time period that expired early in 1916. This seemed to be good news for Ross.

Ross wrote to the Department of Agriculture stating that Seattle City Light was ready to build and that granting a permit to a municipally owned utility was in the public interest. The Department reacted by extending Stone & Webster's permit for a year (undoubtedly with strong persuasion by the company). Ross would not be dissuaded. Upon the second expiration of the permit, he personally approached David F. Houston, the Secretary of Agriculture in Washington, DC, with City Light's permit application in hand. He pointed out that Stone & Webster representatives were buying up other sites while attempting to hold the Skagit without developing it, thereby staking out more sites than Stone & Webster would ever need. The Secretary of Agriculture decided in Ross's favor on December 25, 1918. Ross triumphantly thereafter often characterized Stone & Webster's "greed" as "a case like the boy in Aesop's fable who put his hand in a jar of nuts and in trying to take them all was forced to drop them all."[3]

Stone & Webster continued its attempts to stifle City Light's Skagit project. Company operatives reminded Seattle City Council members of the Cedar River dam fiasco as the Council deliberated approving the Skagit project. When Seattle went to market to sell utility bonds to finance construction, Puget Power pulled strings with federal regulators in an unsuccessful attempt

---

[3] J.D. Ross, "City Light: The Municipal Light and Power System of Seattle, Washington," *Public Ownership,* Vol. X, No. 10, October 1928, p. 181.

to block the sale. With President Coolidge ceremonially pressing a golden key in the White House in September 1924, electricity from the Gorge Dam powerhouse began flowing to Seattle.

J.D. Ross thought big. Gorge Dam was merely a stepping-stone for him. His reputation restored, he envisioned making City Light the sole supplier of electricity for Seattle—and the Skagit (and himself) a showcase for municipal power. He proclaimed that:

> By its very nature the handling of light and power is not in any way a legitimate private business but is a proper governmental function—a monopoly that should belong to the people only .... It is left to us to expand city ownership of light and power and interconnect cities in a superpower system as the most feasible and practical method.[4]

Between 1924 and 1961, the Upper Skagit project was expanded to include a series of three dams, the tallest of which was the 540-foot-tall Ross Dam.[5] The combined generating capacity of the three is more than 700 MW. Ross remained superintendent of City Light until his death in 1939. Four years earlier, in August 1935, when President Franklin D. Roosevelt signed the Public Utility Holding Company Act, which was to be administered by the Securities and Exchange Commission, he appointed Ross an SEC commissioner. In 1937, Roosevelt named Ross the first administrator of the Bonneville Power Administration,

---

4 *Ibid.*, p. 187.

5 The third dam was the Diablo Dam, four miles upstream from Gorge Dam. Construction of the 389-foot-tall arch dam was finished in 1930. The Diablo powerhouse, however, was not completed until 1936. Ross Dam is 5 miles upstream from Diablo Dam. Ross Dam was constructed 1937–49; the first of its four generators began operating in 1952.

which was formed to market Bonneville Dam's hydroelectricity. In 1951, Seattle City Light acquired the Seattle assets of Puget Power. Seattle at last had a unified power system.

Seattle and Tacoma had been both neighbors and spirited competitors from their earliest days—a situation reminiscent of Minnesota's twin cities of Minneapolis and St. Paul. Each was an early member of the municipal power community. In 1873, Tacoma, then a small outpost on the bluffs overlooking Puget Sound's Commencement Bay with Mount Rainier visible in the background to the southeast, had its future assured when the Northern Pacific Railroad—to Seattle's horror—selected it as the future Pacific Coast terminus for Washington's first transcontinental railroad. The transcontinental link became a reality ten years later.

Seattle's growth, however, would permanently eclipse Tacoma's (see Figure 10.2) after Seattle became the transcontinental terminus of the Great Northern Railroad in 1893 and the primary point of departure for the Klondike Gold Rush of 1897–99.

	1870	1880	1890	1900	1910	1920	1930
Seattle	1,107	3,533	42,837	80,671	237,194	315,312	365,583
Tacoma	73	1,098	36,006	37,714	83,743	96,965	106,817

Figure 10.2: Seattle and Tacoma Population History (US Census Data)

In its early days, Tacoma was considered a company town of the Northern Pacific Railroad. The railroad and its officers controlled land sales and development. In 1884, Philadelphia financier Charles B. Wright, a Northern Pacific director and its largest shareholder, obtained a franchise from the city to organize the Tacoma Light and Water Company. The company built a primitive water system that drew water from several creeks and distributed it through pipes made from hollowed-out logs.

Starting in 1885, the waterflow also was used to power an Edison dynamo, and electricity became available to power streetlights and for sale to consumers. This was only three years after Edison's Pearl Street station had become operational on the other side of the continent.

In 1893, Wright sold the water and electrical utility systems to the city, almost a decade before the formation of Seattle City Light. As the city grew, the municipal electric utility (much later named Tacoma Power) distributed power purchased from competing private power companies to meet its expanding needs. By 1907, the Tacoma City Council had had enough of seemingly exorbitant prices for electricity purchased from Stone & Webster—and of the brownouts and blackouts created when Stone & Webster throttled availability to give priority to its other needs. Councilmembers decided that Tacoma Power should build its own hydropower facility and so put a bond measure for the required expenditure to a popular vote. Stone & Webster, in a highly questionable move to thwart the city's plans, announced rate increases and cut off power to the pumps that supplied the city's water.

Irate citizens approved the bond measure by a three-to-one margin in 1909. The LaGrande powerhouse became operational three years later 35 miles from the city on the Nisqually River. A 35-foot-high diversion dam directed water into a settling channel and then into a 2-mile tunnel. The tunnel fed penstocks that dropped 410 feet to the powerhouse.

The LaGrande facility met Tacoma's power needs until World War I escalated requirements. Stone & Webster said it would not sell power to Tacoma Power unless Tacoma agreed never to build additional hydroelectric facilities. Tacoma Power refused. It was impractical to expand the LaGrande project. For a new facility, Tacoma Power selected another site 44 miles northwest of the city on the north fork of the Skokomish River at Lake Cushman, originally a long, narrow broadening of a river formed in a glacial trough and dammed by a terminal moraine from the last ice age. The lake would be an unusually large storage facility in an area known for its heavy rainfall.

Attempts to develop a power plant at Lake Cushman actually first had been initiated by Seattle City Light in 1912, when Seattle citizens approved purchase of the site. As Ross later recounted, "The opposition of [Stone & Webster] delayed proceedings, blocked the city in its hydroelectric development in 1917, and resulted in the loss of the Cushman site."[6] Ross even accused Stone & Webster of planting a hidden microphone in his house to anticipate his every move.[7] Tacoma Power applied for Lake Cushman water rights and reservoir permits in 1919 and began land condemnation proceedings the same year.

After extended and acrimonious property acquisition proceedings, other obstacles, and Stone & Webster roadblocks, Cushman Dam No. 1 construction began in 1924. The constant-angle arch concrete dam was 275 feet tall and 1,111 feet long. The facility, with 43 MW capacity, became operational two years later. The power transmission line to Tacoma stretched across the Narrows between towers more than 1.25 miles apart, the longest single span in the world. Cushman Dam No. 2 was completed in 1930 just downstream from Dam No. 1.

---

6 J.D. Ross, "Seattle City Light and Power," *Public Ownership of Public Utilities*, Vol. XVI, No. 5, May 1934, p. 84.

7 Paul Dorpat and Genevieve McCoy, *Building Washington: A History of Washington State Public Works*, Seattle, WA: Tartu Publications, 1998, p. 284.

The Cushman dams were designed by Lars Jorgensen, a noted California-based engineer who had designed the world's first constant-angle arch dam: Salmon Creek Dam built near Juneau, Alaska, in 1913.[8] The Cushman Dams were so successful that J.D. Ross, in an attempt to counteract criticism of his botched Cedar River upgrade, engaged Jorgensen to design the Diablo Dam, which he constructed on the Skagit River after the Gorge facility was completed.

In 1924, Seattle City Light and Tacoma Power built a tie line so that the two municipally owned utilities could share power when necessary. As Ross quipped, "The results to the two cities may be expressed by saying 'blest be the tie that binds.'"[9]

While Stone & Webster was doing everything it could to restrain the municipal power movement in Seattle and Tacoma, it simultaneously was following Samuel Insull's model to refine and extend its electrical grid in western Washington. Load growth throughout the company's system, particularly in the northern portion, required power development there. The only hydroelectric plant operated by the company in that region was the tiny, outdated Nooksack Falls plant (1,750 kW). For many years, Puget Sound Power and Light had been forced to divert power from its southern generating plants in the Seattle-Tacoma area and to purchase power from Canada to adequately serve the

---

[8] Jorgensen developed the theory for constant-angle arch dams. It was estimated that these dams required 20 percent less concrete than constant-radius arch dams that had been the norm. Both of these thin arch designs required significantly less concrete than traditional gravity dams. In addition to single arch dams, Jorgensen designed multiple-arch dams such as Gem Lake Dam on the eastern slope of the Sierra Nevadas in California. Reclamation consulted with him during its initial planning for Hoover Dam.

[9] Ross, "City Light," p. 180.

Bellingham area. Bellingham itself, located on Bellingham Bay about 20 miles south of the Canadian border and 30 miles west of Mount Baker, was growing rapidly: Its population jumped from 11,000 in 1900 to 26,000 in 1920. In 1918, the load on the company's entire system began exceeding the capacity of its complement of waterpower plants.

Puget Sound Power and Light decided to build a new hydroelectric plant near Mount Baker on the Baker River just north of the town of Concrete. This would allow the company to better serve the northern areas of its system, rebalance its entire power grid, and discontinue purchasing electricity from Canada.

From near the Canadian border, the Baker River flows southward 30 miles through a steep glacial valley near Mount Shuksan and Mount Baker before joining the Skagit River at the cement-manufacturing town of Concrete. The river is fed by glaciers, heavy rains, and snow fall. From Concrete, the Skagit flows west about 25 miles into Puget Sound. Seattle City Light's Gorge Dam is approximately 40 miles up the Skagit from Concrete.

For the last half of its run, the Baker River traverses a nearly level valley closed at its southern end by a natural limestone wall more than 500 feet high and 2,000 feet thick. The river flows through Eden Canyon, a narrow gorge in this barrier with nearly vertical sides, then downstream 1.1 miles through Concrete and into the Skagit River. The canyon riverbed is roughened, solid limestone. Clearly, Eden Canyon was geologically an ideal site to locate Baker Dam. Furthermore, cement was available locally at Concrete, and sand, gravel, and stone were close at hand. The dam site could be connected to the Great Northern Railway by laying half a mile of track.

The Baker River site had been purchased in 1915 to be held in reserve until Stone & Webster's power needs would justify its development. The purchase undoubtedly was part of the company's efforts to impede Seattle City Light. Circa 1921, Stone & Webster's western regional staff in Seattle began developing, under the supervision of W.D. Shannon, detailed plans for the Baker River project. This included dam and electrical generation and transmission system design, construction plans, cost estimates, permitting, and myriad other details. Shannon, a University of Michigan civil engineer, also simultaneously oversaw a major upgrade (from 44 MW to 60 MW) of the company's White River hydroelectric facility just east of Tacoma. Later the Baker Dam reservoir would be named Lake Shannon in his honor. He went on to become a prominent Seattle citizen and a state senator.

Figure 10.3 provides an overview of the project concept that emerged.

Figure 10.3: The Baker River Project

Baker Dam is a 293-foot-tall, arch-gravity structure with spillway through control gates over the dam crest. The upstream face follows a 250-foot radius. The 1,600-foot-long diversion tunnel, bored through solid limestone, is 22 feet in diameter and lined with concrete. With a height almost identical to the length of a football field, the dam was the highest hydroelectric dam in the United States when completed. Its initial power output was 40 MW. After several upgrades, it now is 111 MW. The reservoir behind the dam, Lake Shannon, extends upriver 8 miles to another dam built in 1959. That dam now is called Upper Baker Dam; the original Baker Dam is referred to as Lower Baker Dam. Upper Baker Dam generates 107 MW of power. The dams are managed together to provide flood control for the Skagit basin downstream. These significant hydroelectric facilities continue to be key components of the electric grid operated by Puget Sound Energy, the successor company to Puget Sound Power and Light.

Construction of Baker River project preliminaries such as worker housing and a railway spur connecting to the Great Northern Railway line nearby began in April 1924. At about the same time, while George Jessup was completing the construction of the Henry Ford hydroelectric facility in Iron Mountain, Michigan (see Chapter Nine), he was told by Stone & Webster senior management that his next assignment was to be superintendent of construction for the Baker River project and that he was to report to Concrete as soon as possible.

Jessup clearly was a rising star at Stone & Webster, but his selection was not an obvious one. The Concrete project was much more complex than Iron Mountain—Jessup's only other dam-building experience. Nevertheless, he quickly boarded a train headed west. His wife and their three young children, an infant son and daughters ages five and nearly three, soon followed from Iron Mountain once family quarters were built in the construction camp on the bluffs above the dam site.

Construction proceeded at breakneck speed. By late July 1924, the railroad to the dam site had been completed, the construction camp had become a small city, and more than 900 men were transporting materials, building roads, erecting buildings, clearing ground, blasting rock, and doing the many, many other things that were necessary before actual construction work on the dam and powerhouse could begin. In August, the diversion tunnel bored in the canyon sidewall was completed, and the river's water began flowing through it to bypass the dam site. Coffer dams were built upstream and downstream of the dam site, and the riverbed was excavated down to the bedrock that was to support the dam's foundation.

Heavy rainstorms in September 1924 led to an unexpected rise in the river, washed out trestles and other structures on the dam site, inundated the area between the coffer dams, submerged machinery and equipment, and delayed operations by several weeks. Another rainstorm early in October again overflowed the coffer dams.

On October 10, the first concrete for the dam's foundation was poured. A week later, workers represented by the Industrial Workers of the World went on strike demanding "a 25 percent wage increase, more and better food, clean linen once a week, no overtime, safer working conditions, boycott of California products, and release of all class prisoners."[10] As the strike wore on, it became violent. In an incident seared in memory for the

---

[10] *The Concrete Herald*, Vol. XXIII, No. 47, October 23, 1924, p. 1. Members of the IWW typically were referred to as "Wobblies." It was not shown how Stone & Webster could comply with the last demand.

rest of their lives, Jessup's two young daughters were awakened one night by a noise at the window of the bedroom they shared. They started screaming upon seeing a man looking in and brandishing a gun. Jessup family members were placed under armed guard until the strike ended. At one point, as the strike continued, a doctor had to be escorted by armed security men as he walked along the railroad tracks from town to Jessup's quarters to check on Jessup's ill infant son. According to Jessup, the strike was settled when each worker was given a bedsheet.

The timing of the strike was unfortunate. With a few more days' work, enough concrete would have been poured to protect the project from high water. The 1924–25 winter season was one of the wettest on record, with monthly rainfall of around 15 inches through February, causing wave after wave of flooding. It was not until March 1, 1925, that work crews were able to pour concrete to the height of the upstream coffer dam and entirely shut off the river flow.

In Summer 1925, about thirteen hundred men were working around the clock to complete the project. There was a shortage of at least three hundred men due to an urgent demand for men to fight forest fires raging in the area. One day, Jessup fell from the bluff above the river into the canyon below, breaking several ribs and sustaining multiple serious injuries. When workers brought him to his quarters motionless on a stretcher, his wife initially thought he was dead. Nonetheless, he somehow was able to recover quickly enough to resume managing the project construction. A Stone & Webster photograph (see Figure 10.4) taken a few months before the dam was completed shows Jessup (far left), normally a robust man, standing hunched over and emaciated:

Figure 10.4: Stone & Webster Officers at Baker River
Far left, George Jessup; fourth from left, Edwin Webster; fifth from left, W.D. Shannon; second from right, Charles Stone; far right, Samuel Shuffleton

Remarkably, especially given the impediments overcome during construction, the Baker project was completed ahead of schedule and in record-setting time. In October 1925, dam and powerhouse construction was completed, and Lake Shannon began to rise behind the dam. A month later, the hydropower facility became operational. Stone & Webster claimed the project construction set a world's record: No other plant of equal power was known to have been constructed in as short a period of time.

While the Baker dam and powerhouse were being built, Stone & Webster simultaneously was constructing 92 miles of high-voltage transmission lines and two large substations. Once the Baker River plant became operational, power generated there could be delivered to almost any part of Washington served by Puget Sound Power and Light.

Immediately after the Baker River facility began transmitting electricity, Puget Sound Power and Light placed full-page advertisements in Seattle-area newspapers to continue its feud with

J.D. Ross and Seattle City Light. Under the banner headline "More 'Puget Power,'" the advertisements proclaimed:

> Baker River is harnessed to add its age-old strength to the upbuilding of the Pacific Northwest. … The investment of over eight thousand citizens of Western Washington in our securities has aided in making this plant possible. It will be added to the taxable wealth of this state and will aid in reducing the taxes not only of the people of Skagit County, where the plant is located, but of every taxpayer in the state. Just a few miles away is the plant of the City of Seattle, tax-exempt and tax-free. The Baker River plant will not only light the homes and stores of the Pacific Northwest, but will furnish the power for new factories, new industries, adding more payrolls and more taxable property to the entire Puget Sound District. It is another step in the industrial progress of this state—additional proof that the Puget Sound Power & Light Company will always maintain an adequate supply of electric power well in advance of the needs of the district which it serves.[11]

There is an interesting tall tale, an important fish story, associated with Baker Dam. Hydropower developers of that era were notoriously insensitive to environmental issues. That was not the case with the Baker River project. At the time the dam was to be constructed, the Baker River was the only stream in Washington State in which sockeye salmon spawned. The sockeye salmon, 24 to 33 inches long and weighing 5 to 15 pounds, is among the smaller of the seven Pacific salmon species, but

[11] This text appeared in an advertisement in the *Seattle Post-Intelligencer,* November 27, 1925, and in the Sedro-Woolley *Courier-Times,* November 26, 1925.

its succulent, bright-orange meat is highly prized. As with all other Pacific salmon, sockeye journey upriver from the ocean to spawn in fresh water. They require a nearby lake in which to rear their offspring. Once hatched, juvenile sockeye stay in their natal habitat for one to two years. They then journey out to sea, where they grow rapidly, feeding mainly on zooplankton. They stay in the ocean for one to four years. It was critical to the Pacific Northwest's fishing industry, to the livelihood of Native American tribes living in the area, and to nature's ecological balance more generally that the Baker Dam allow passage for adult salmon migrating upstream to spawn and for salmon fry heading downstream toward the ocean.

But the dam's 293-foot height presented a conundrum previously never encountered. Fish never before had been lifted over an obstruction more than 50 feet high. Project managers consulted with state and federal fisheries officials, area Native American tribes, and renowned marine biologists and university researchers. A committee was formed to address the various issues involved. After a number of committee meetings, it was evident that there were a number of conflicting ideas and that no resolution was imminent. Jessup later related that he told the committee he had a dam to build and a schedule for completing it. Based on all he had heard, he was going to proceed with the design and construction of a Baker fishway, working directly with the state superintendent of hatcheries. The result was a unique and highly innovative lift system.

It incorporated a forebay downstream of the dam adjacent to the tailwaters of the powerhouse, where salmon were corralled and confined into a fish ladder with 2-foot falls. The fish ladder was designed so that the fish could move upwards from one pool to the next but could not return. The 2-foot elevation between pools was chosen so that the fish could conserve strength between jumps. The fish ladder reached upward to a staging area partway up the height of the dam. There, the fish entered a water-filled car and were transported on an incline railway the rest of the way to the top. In addition, one of the dam's spillway gates was left open during the June run of fry to the ocean. The danger that fry would encounter passing over the dam on their way downstream was mitigated by designing the dam spillway with a special apron at its base to spread the fall of the water and smooth the transition to the river downstream. It also was discovered that most fry that happened to enter the powerhouse water intake tunnel were able to pass through the powerhouse turbines unharmed.

According to press reports, the success of the Baker fish transport system was lauded as a major advance and studied by the fish management community and power plant engineers in all parts of the world. The *Journal of Electricity* reported that:

> The success of the whole enterprise means a great deal to both the salmon and the power industry of not only this state but of the whole country, fisheries experts have declared. This is the first time so far as is known that the migratory fish have been successfully transported over a high dam. It is predicted that no longer will the power companies be restrained from building as high a dam as is needed across any of our salmon streams and no longer will the great salmon industry of the state be menaced as a result of such power dams.[12]

This innovative approach to solving the salmon issue was just one of the ways in which western hydroelectric facilities differed

[12] *Journal of Electricity*, Vol. 57, No. 6, September 15, 1926, p. 197.

from their eastern brethren. First, western dams usually are much taller (i.e., higher head) since they can be placed where waterways run through draws or canyons. Second, due to their remote location, western dams often require power transmission over long distances. Third, construction in remote, rugged, western locations can be especially difficult. Fourth, seasonal water flow variability and topographic specifics can lead to reservoir storage far from dams and/or power plants and connection via lengthy flumes or tunnels.

As Washington's electrical demand continued to increase after the completion of Baker Dam, both Puget Sound Power and Light and Stone & Webster zeroed-in on Rock Island on the Columbia River as an additional hydropower site. The site was 12 miles downstream from Wenatchee and 463 miles upstream from the mouth of the river.

Four years after Baker became operational, the Federal Power Commission issued Puget Sound Power a license to build a Rock Island hydroelectric facility, the first to span the Columbia. Despite the onset of the Great Depression, construction began in January 1930. The 78 MW facility became operational in February 1933. Building upon the lessons of the Baker River project, a gently sloping fish ladder was included. After several major upgrades, the facility now is rated at 624 MW. In 1956, the Rock Island facility was acquired by the Chelan County Public Utility District, which continues to operate it.

The investor-owned Rock Island Dam was not a multi-purpose dam: Its sole purpose was to generate electricity. In the same year that Rock Island became operational, construction of the mammoth, federally funded, multiple-purpose Bonneville and Grand Coulee dams began downstream and upstream on the Columbia.

The battle between publicly owned and investor-owned power utilities shifted dramatically within the decade after the three dams highlighted in this chapter were completed. The Great Depression, President Franklin D. Roosevelt's New Deal, the dismantling of utility holding companies, the advent of large, multipurpose federal hydropower projects, and other factors altered the playing field.

Undertaking new hydroelectric projects became increasingly difficult, especially for investor-owned utilities. In Washington State, voters in 1930 approved legislation allowing the formation of county public utility districts for electricity distribution. These PUDs could acquire properties of investor-owned companies by condemnation. Investor-owned Puget Power and Light and Stone & Webster had fought hard against the bill.

By 1936, thirteen Washington counties had formed PUDs. Then, in 1937, Congress created the Bonneville Power Administration to distribute the cheap and abundant electricity to be generated by the Bonneville and Grand Coulee dams. Prices were to be the same for all users, and publicly owned utilities were to be given preference. With service areas and revenue bases threatened, investor-owned companies found financing new construction projects to be extremely difficult. Hydropower's entrepreneurial days were ending.

Fig. 10.5. Seattle City Light's J.D. Ross

Fig. 10.6. Seattle City Light's Gorge Dam Powerhouse
(left section added in the 1940s)

Fig. 10.7. Construction of Seattle City Light's
Cedar River Dam Nearing Completion

Fig. 10.8. Tacoma Power's Cushman No. 1 Dam and Power House

Fig. 10.9. Tacoma Power's Record-Setting 1.3 Mile Single-Span Transmission Line from Cushman No. 1

Fig. 10.10. Baker River Dam Site Before Construction Began

Fig. 10.11. Puget Sound Power & Light's Baker Dam

W.D. Shannon
Genl. Supt.

G.P. Jessup
Supt. of Const.

Fig. 10.12. Stone & Webster Project Managers At Baker River 1925

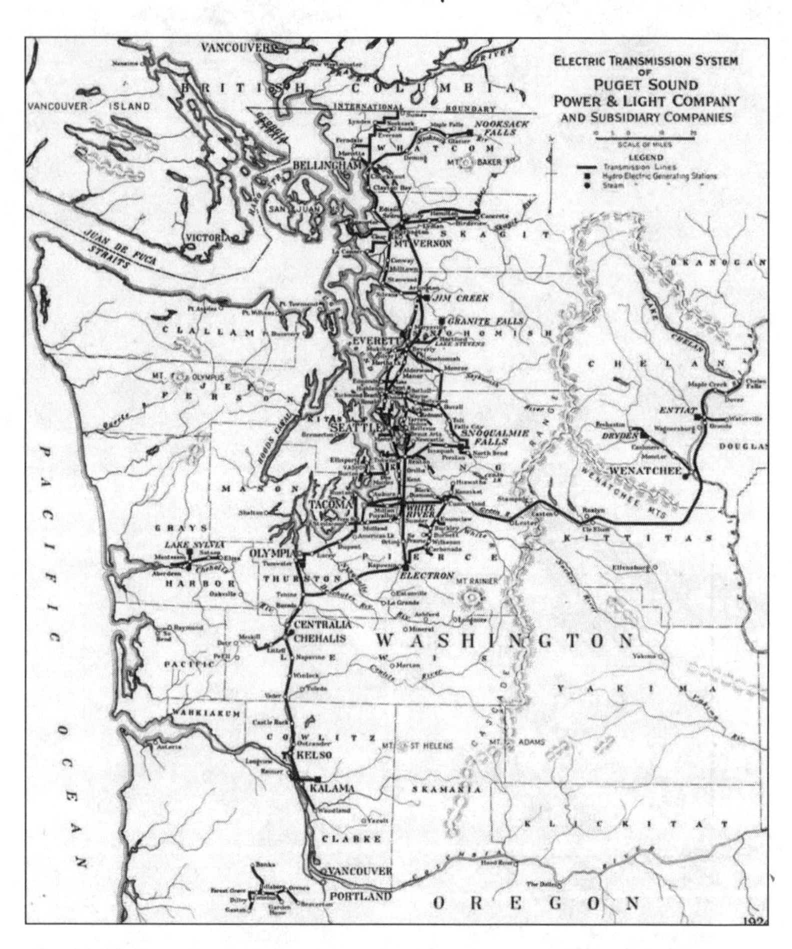

Fig. 10.13. Puget Sound Power & Light's Electric Transmission System in 1924–25

Chapter Eleven

Conowingo

As our car passed Havre de Grace in Maryland at the start of a family vacation when I was a young boy, my mother pointed westward and said, "There's your granddaddy's dam." For a moment, faintly in the distance, I saw what looked like a railroad bridge spanning the Susquehanna River. Some years later, I passed that way again and remembered what my mother had said. Suddenly curious, I looked at a map and learned the dam was Conowingo Dam. I detoured and, 10 miles later, encountered the dam up close. Holy cow! What a sight! I had to know more.

Unlike the Mississippi River or the Colorado River, most people know little, if anything, about the Susquehanna River. Yet it is the largest river flowing entirely within the United States that drains into the Atlantic Ocean. It rises in Cooperstown, New York, and, over its 464-mile run, drains large portions of the southern tier of New York and nearly half the land area of Pennsylvania before emptying into the Chesapeake Bay at Havre de

Grace. Pennsylvania cities along its course include Sunbury, Williamsport, Harrisburg, and Wilkes-Barre/Scranton. Due to its rocky, shallow nature, it is the longest, commercially non-navigable river in the United States. It is highly flood prone, experiencing a devastating flood on average every fourteen years.

The Susquehanna gradually drops about 150 feet in its last 25 miles. The small town of Conowingo (whose name means "at the rapids" in the language of the Native Americans who originally inhabited the area) overlooked the river just over 10 miles upstream from its mouth. From around the time that the power of Niagara Falls was developed, various schemes were considered to harness the Susquehanna's gigantic hydropower potential near Conowingo. The potential ranked up there with Niagara Falls and Muscle Shoals. The situation, however, was complicated by the river flow's wide swings. Average annual flow varied greatly, and seasonal and unexpected variations from annual averages were large.

Conowingo is about 70 miles southwest from Philadelphia, home to Philadelphia Electric Company, which evolved from an original Edison franchise granted just four years after Edison's Pearl Street Station had become operational in Manhattan. PECO enjoyed a service monopoly in Philadelphia after 1902 and was the nation's seventh largest utility by the early 1920s. It was the logical candidate to develop Conowingo's potential.

From the time PECO was formed, it used the abundant coal mined in Pennsylvania to fuel steam power plants to meet its demands for electricity. Thermal plant efficiencies steadily improved from economies of scale and improvements in boiler, turbine, and generator design. It was estimated that plants brought on line in the mid-1920s required one fifth as much fuel as their counterparts twenty-five years earlier in order to produce the same amount of electricity. Even though PECO continued to upgrade its steam plants and build much larger new ones, demand was outstripping supply by World War I. PECO faced other major issues as well. Among them were disruptions in the supply of coal and escalating coal prices. After the war, in 1919 and again in 1922, strikes shut coal mines. Moreover, there were persistent shortages of rail cars to transport coal. Given the situation, significant additional generating capacity and system reliability became crucial for PECO.

During World War I, the federal government considered the Conowingo area on the Susquehanna during its search for additional power, discussing the possibilities with PECO. The proposition was dropped due to the potentially long development time, but the idea continued to resonate with PECO.

Beginning in 1921, William C.L. Eglin, PECO's vice president and chief engineer, initiated preliminary studies to assess developing a Lower Susquehanna hydropower facility. Eglin was an extraordinary, highly capable individual. Born in 1870 in Glasgow, Scotland, he attended university there and even took special classes with Lord William Thomson Kelvin, one of the preeminent scientists of his day. Upon coming to the United States in 1889, Eglin joined the Edison Electric Light Company of Philadelphia, which, over time, became PECO. He rapidly advanced at Edison Electric/PECO and became a widely known and respected authority on electric lighting. In 1908–09, he served as president of the National Electric Light Association, through which he interacted extensively with Samuel Insull and presidents of the country's other leading electric utilities. He became vice president of the American Institute of Electrical Engineers and long-time president of the Franklin Institute. He received honorary doctoral degrees from the University of Pennsylvania and Swarthmore

College. He was described as "testy, impulsive, and very sure of himself" and "an aggressive, brilliant worker."[1]

Eglin also was an experienced power plant designer and project manager. From 1902 through 1921, William Windrim, the renowned Philadelphia architect famous for his neoclassical designs, teamed up with Eglin to plan, design, and oversee construction of five spectacular PECO coal-fired central power stations. PECO contracted with Stone & Webster for design services and as general contractor for construction of the latter two of these plants: the Delaware Station (operational in 1920; 183 MW) and the Richmond Station (operational in 1925; 100 MW). Given Stone & Webster's preeminent position in the electric industry and the fact that the company had built and run Philadelphia's mammoth Hog Island Shipyard, to which PECO supplied electricity, its selection was not surprising. Stone & Webster also had managed the initial public offering of PECO stock in 1903.

When Eglin began his Conowingo studies, he again engaged Stone &Webster for assistance. The enormous power potential at Conowingo and the wide variability in the river's flow made the development of a hydropower facility at Conowingo to transmit power to nearby Philadelphia especially attractive. When the Susquehanna was running strongly, the electricity it generated could carry PECO's base load, and PECO's coal-driven plants could handle peak demands. Conversely, when the river was low, PECO's steam plants could satisfy base load, and Conowingo could handle peaks. A Conowingo hydropower plant would be ideal for supplying peaking power. Whereas thermal plants needed time to build steam pressure, hydroelectric plants could start generating power in minutes.

---

1 Hughes, *Networks of Power*, p. 329.

A primary first concern was site availability. Companies under the control of investment bankers Bertron, Griscom & Company owned more than 5,000 acres of land around Conowingo and held hydroelectric rights on the Lower Susquehanna. Bertron, Griscom offered PECO an option on the land and hydroelectric rights. In 1923, Eglin *et al.* presented their preliminary business case for Conowingo development to PECO's Board of Directors. They concluded that it was economically feasible to build a Conowingo dam and 250 MW power plant and to transmit power to Philadelphia at a lower cost than if it were produced by PECO's steam plants. But there was a snag. The terms of the Bertron, Griscom option were onerous, requiring the sale of 25 percent of Conowingo's output to another utility at below-cost. To keep the project prospects alive, subsequent to side discussions between Eglin and a representative of the Philadelphia investment banking house of Drexel & Company who served on the PECO Board, Drexel purchased the option to protect the company and renegotiated its terms.

Additional engineering and financial studies were conducted. These more detailed studies concluded that, although the projected project budget was $59 million (over $850 million in today's dollars), it also would:

- cost $21 million less than building steam plants of equivalent power,
- save 750,000 tons of coal annually,
- allow 36 MW of power to go into emergency service in a matter of minutes,
- provide hydroelectric power at times of the year when coal deliveries were most troublesome, and
- enhance overall system reliability.

In July 1924, PECO's Board advised Drexel that the company would exercise the site option and authorized preliminary work at Conowingo.

Yet regulatory and financial hurdles remained. The dam would be in Maryland, and the upstream reservoir (or, pond) would back up into Pennsylvania. Thus, it was necessary to obtain project approval from the public service commissions of both states. Because the Susquehanna had been ruled a navigable river by the War Department,[2] it also was necessary to obtain a license from the Federal Power Commission. A convoluted organizational structure was devised to accommodate the jurisdiction and requirements of the three commissions, and detailed cost estimates and engineering and financing plans were prepared. A joint hearing of the three commissions was held in March 1925. Acrimonious opposition—led by PECO's largest customer, which was making a blatant hostile attempt to acquire PECO—surfaced immediately and continued for another year before PECO finally received its necessary approvals. A syndicate formed by Drexel then marketed a $48 million PECO bond and preferred stock financing, and the project entered its construction phase.

The project was immense and complex. When completed, it would be the second largest hydroelectric plant in the world, behind the Sir Adam Beck #1 plant completed in 1922 on the Canadian side of Niagara Falls. By the time of its completion, the PECO Conowingo project would be the largest development—steam or hydro—constructed in one step as a single project in the history of the power industry.

---

[2] Nonetheless, the Susquehanna River remains commercially non-navigable today.

The dam site was chosen to maximize available head and because the steep rocky bluffs walling the riverbed on both sides there would form natural dam abutments. Unfortunately, however, the reservoir created behind the dam would submerge the village of Conowingo 2 miles upstream. More than two hundred inhabitants would have to evacuate before the village was demolished.

The Baltimore Pike (Route 1) crossed the river on a bridge in the village as well. It would be rerouted to cross the river atop the dam. In addition, the tracks of the Columbia & Port Deposit branch of the Pennsylvania Railroad, an important freight route, ran along the eastern bank of the river (the bank to the right facing the dam from its downstream side) and would need to be relocated to higher ground over a nearly 20-mile stretch.

The concrete gravity dam's total length is 4,648 feet (0.9 mile). It rests on solid granite. The dam's height to the relocated Route 1 roadway topping it is 105 feet. The powerhouse is located at the western end of the dam, where the river channel is deepest. The section of the dam just east of the powerhouse contains three regulating gates used frequently to control and fine tune the elevation of the water in the reservoir (aka Conowingo Pond). Farther east are fifty spillway sections topped by 42-ton steel flood gates that are raised or lowered using three gantry cranes running on tracks beside the highway. The same gantry cranes operate the regulating gates. The remaining portion of the dam to the east is a solid, buttressed wall that continues the highway roadbed. The 9,000-acre Conowingo Pond behind the dam extends approximately 14 miles upstream.[3] The original

[3] The Holtwood hydroelectric facility (107 MW) had been built upstream from the Conowingo dam site by another electric utility and had been supplying power to Baltimore since 1911. The Conowingo Dam was de-

dam design had included fishways, but, at government request, they were eliminated and PECO instead paid an annual fee for stocking Conowingo Pond with fish.

The power plant is 950 feet long—longer than three football fields placed end to end. Its upstream east–west wall is a structural continuation of the dam. Its downstream face features a series of sixteen large, dramatic, arched windows to illuminate the huge generator room. The floor of the generator room is about six stories above the base of the dam, and the room itself is about six stories tall. Perched spectacularly atop the roof is an immense high-voltage switchyard connecting the plant's generators to each other and to 220,000-volt transmission lines sending power to Philadelphia.

The power plant initially housed seven 36 MW generators (for a total of 252 MW) but was built to provide for future installation of four additional generators. The facility's installed capacity now is 575 MW. Water from Conowingo Pond flows through penstock intakes toward the water wheels (turbines), which are shaft-mounted below the generators. The water flow to each turbine is controlled by a butterfly valve installed at the entrance to the turbine scroll case. These 27-foot-diameter valves, the largest ever made at the time, each weighing 136 tons, were designed to open or close in five minutes against maximum water flow. The turbines and generators were the largest ever made at the time.

PECO engaged Stone & Webster for the design and construction of the Conowingo hydroelectric facility. Stone &

---

signed to ensure that its reservoir would not impinge upon Holtwood's tailwaters. Hugh L. Cooper designed and initially oversaw the construction of Holtwood Dam. He subsequently designed Keokuk and Wilson dams (see Chapters Seven and Eight).

Webster itself built the powerhouse and supervised construction by the Arundel Corporation of Baltimore of the remaining eastern portion of the dam and relocation of the Columbia & Port Deposit railroad tracks. The transmission system was designed and constructed by Day & Zimmermann. My grandfather, the same George Jessup that we learned about earlier, was Stone & Webster's superintendent of construction for the power plant.

Construction of the dam and power plant began in March 1926. A temporary village (complete with sewage and water system, hospital, mess halls, and police force) was built on each side of the river to house a labor force of 3,800 men. A 10-mile railroad spur was constructed to the dam site to carry men, equipment, and materials from Havre de Grace. Over time, it handled 13,000 cars and carried 500,000 passengers.

The construction effort was mind boggling. Even though hydropower facility construction was dangerous work, particularly in this pre-OSHA era, men of many nationalities and ethnic backgrounds swarmed to the area seeking the project's high-paying jobs (35 to 60 cents per hour, depending upon skill level). At the peak of activity, nearly 5,500 men were at work. Roughly 800,000 tons of rocks were excavated from the granite riverbed to form the foundation for the dam and powerhouse. Nearly 8 million feet of timber were used building coffer dams. Almost 700,000 cubic yards of concrete were poured, with an incredible 58,000 cubic yards poured in one month alone. Although no official records were kept of injuries or deaths during construction, the deaths of sixteen laborers have been confirmed.

The pace of work was furious: All parties involved wanted coffer dams in place before seasonal river flooding might destroy them. Those fears were well founded. The worst Susquehanna

River flood in thirty-five years came within 3 inches of breaching the coffers in November 1926.[4]

Relocation of the Columbia & Port Deposit railroad tracks along the eastern bank of the river was quite a project in its own right. It required blasting 900 feet of tunnels through solid granite, erecting new bridges, and laying nearly 20 miles of tracks. Most of the trackage was benched into the rocky hillsides.

Meanwhile, Day & Zimmermann was constructing the high-voltage transmission lines to Philadelphia. Two lines of steel towers carried 220 kV electricity 58 miles to a substation constructed at Plymouth Meeting, about 15 miles northwest of downtown Philadelphia. Here the voltage was stepped down to 66 kV and the power conveyed into Philadelphia on towers built alongside and over the tracks of the Reading Railroad. In total, more than a thousand transmission towers were erected.

On January 18, 1928, the gates of the dam were closed, and the reservoir behind the dam began to fill. Two miles upstream, displaced residents of the village of Conowingo watched emotionally as their homes disappeared. By late evening, the entire village was completely submerged. The growing reservoir also inundated numerous petroglyphs carved into rocks along the river by Native Americans 500 to 1,000 years ago. At the time Conowingo was constructed, scant attention was paid to the human and cultural impact of dam projects. That would change over decades to come.

The entire construction project was completed in less than two years—six months ahead of schedule. Commercial opera-

---

4 A riveting silent movie documenting the construction has been discovered. See footage of the river floodwaters rising and almost breaching the coffers at https://www.youtube.com/watch?v=qShnoumydKI#action=share

tion commenced in March 1928. With final costs of around $52 million ($755 million today), the project also was achieved at well under the originally approved budget of $59 million. It continues to be operated by Exelon, PECO's successor.

William C.L. Eglin, who, more than anyone else, had been responsible for this hydroelectric development, was deprived of the satisfaction of seeing its completion. He died at the age of fifty-seven, just three weeks before Conowingo became operational. He had been stricken four weeks earlier with an unexplained "nervous malady" while on a Caribbean cruise.

Although the vast Conowingo project was one of the greatest hydropower developments of its day, its importance is everlasting. It was the linchpin of the nation's first planned system to regionally integrate independent utilities via a high-voltage transmission network. The utilities, which previously had evolved independently, were knit together by the high-voltage network, preserved their legal identities, pooled their generating capacity, and acted as distributors of the pooled energy. The model has endured, growing into today's national utility grid.

While Eglin and other senior managers at PECO were doing their initial planning for Conowingo, they quickly expanded their thinking to include the possibility of linking with two neighboring utilities: the Public Service Electric & Gas Company of New Jersey and the Pennsylvania Power & Light Company, which operated in central and eastern Pennsylvania. The market for Conowingo power would be expanded, and load and power generation diversity within the combined service area would increase. A primary goal would be to achieve a dynamic balance of generating costs and capacity throughout the system. Eglin observed that, "[o]nly the most alert and skilled load dispatching will realize all of the possible benefits [from interconnection]."[5] Interconnection and integration also would allow the three utilities to increase system reliability by pooling risks while simultaneously reducing aggregate reserve capacity since it was very unlikely that catastrophe or maintenance requirements would strike all three participants simultaneously.

In September 1927, while Conowingo construction was proceeding feverishly, the three utilities signed an agreement formally establishing the PNJ Interconnection as an integrated, centrally controlled pool of electric power. The agreement called for a 210-mile-long ring of 220,000-volt trunk lines, with two transmission lines from each utility's system to the other two, as shown in Figure 11.1 (220,000-volt ring noted by dotted line).

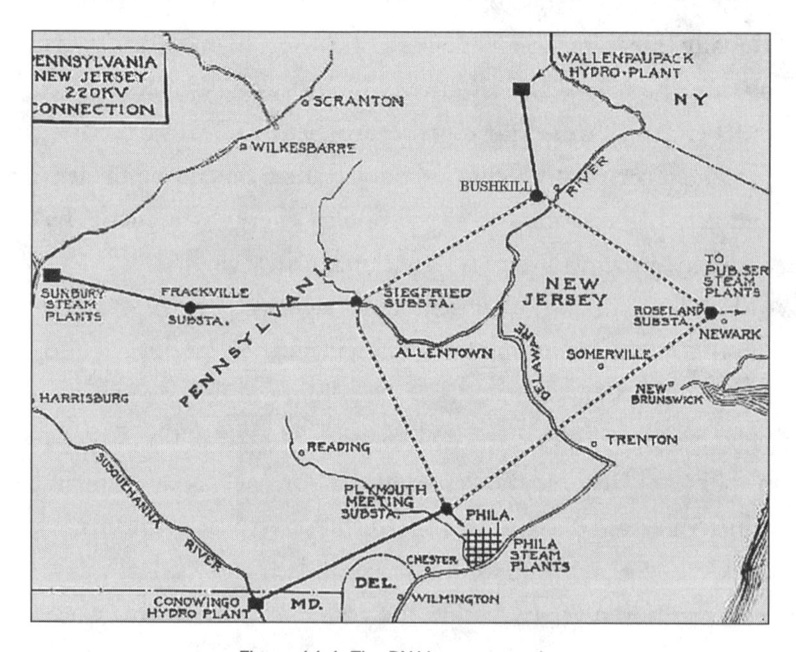

Figure 11.1: The PNJ Interconnection

5 Hughes, *Networks of Power*, p. 332.

Construction responsibilities and costs were allocated among the three utilities. Power from each utility's generating stations fed into the interconnection ring. The utilities also took bulk power from the interconnection substations to meet their needs. They agreed not to tap into the ring directly to service customers. Connections to the high-voltage ring provided access to a range of complementary power sources. For example, hydropower came from Conowingo and from PP&L's Wallenpaupack facility. PP&L contributed power from a thermal plant located at the coal mines near Sunbury.

Load peaks among the three companies varied greatly, and this provided substantial overall load diversity. A control center operated by PECO on behalf of the consortium dispatched the power generated by all three utilities.

A committee of peers from each company negotiated planning, operations, and general financial matters for the interconnection.

The PNJ Interconnection became fully operational in 1930 with a pooled generating capacity of 2,250 MW. By all accounts it was highly successful. Managers and engineers from all over the world came to learn about it. It evolved into today's Pennsylvania–New Jersey–Maryland Interconnection (the PJM grid), with a pooled generating capacity of approximately 180,000 MW that enables delivery of electricity to more than 65 million people in all or part of thirteen states and the District of Columbia.

As part of the PJM grid, Conowingo plays a vital role in the event of a large-scale regional blackout, such as the one that crippled New York City and affected more than 50 million people in eight states and parts of Canada in August 2003. Conowingo can generate power quickly by opening its gates to release water to propel turbines and bring generators online. That power can be used to jump-start other grid power plants that cannot self-start (e.g., coal, nuclear, natural gas).

Conowingo serves as an example of America's reputation for being able to do big things and its can-do attitude. It was publicized extensively upon its completion, hailed as an extraordinary, awe-inspiring feat.

As we saw in Ford's battle for Wilson Dam and again in J. D. Ross's municipal power crusade in Washington State, this was an era of intense debate over the ownership and control of electric utilities. PECO had been hobbled by insufficient generating capacity and crippling power outages since the days of World War I. It was crucial for the company to show that it had brought ample, reliable capacity online to serve the public interest and that it had done so quickly, efficiently, and economically. PECO wanted to convey stability, permanence, and civic responsibility at a time when investor-owned electric utilities faced considerable public scrutiny. Simply put, PECO wanted to make the case that private ownership and public interest were synonymous. Conowingo was entirely privately financed—the federal government did not invest a penny.

PECO advertisements proclaimed that "power means plenty" and proudly pointed to Conowingo as "one of the world's greatest waterpower installations, built, owned, and operated by Philadelphia Electric."[6] Photographs of the dam were stylized to emphasize that it harnessed the power of the mighty Susquehanna. Pictures of the interior of the power plant's generator room highlighted its grandeur. In 1929, a replica of the project was placed in the Smithsonian Institution.

---

[6] John R. and Erin E. Paulson, *Conowingo*, Charleston, SC: Arcadia Publishing, 2017, p. 91.

A future president of PECO said of Conowingo in a speech that:

> It came into being as a product of the foresight and constructive imagination of courageous and pioneering Americans, who provided the necessary money and engineering talents. Conowingo has made its contribution to the economic weal. It has shown how citizens working together can develop such natural resources in the national interest while subject to state and federal regulation. It has provided employment. It has made substantial payments in the form of county, state, and federal taxes. It has made that portion of the Susquehanna valley a living part of our economy. ... Conowingo, built by a free people, for the public good, is truly a great development on the Susquehanna![7]

---

7 R.G. Rincliffe, *An Address to The Newcomen Society in North America*, Philadelphia, PA, November 24, 1953.

Fig. 11.2. W.C.L. Eglin, Philadelphia Electric Company

Fig. 11.3. PECO's Richmond Coal-Fired Central Power Station

Fig. 11.4. Conowingo Dam and Powerhouse

Fig. 11.5. Conowingo Powerhouse

Fig. 11.6. Conowingo Generator Room

Fig. 11.7. 27-Foot Diameter Conowingo Turbine Butterfly Valve

Fig. 11.8. Conowingo Control Room

Fig. 11.9. PECO Transmission Tower over Reading Railroad Right of Way

Fig. 11.10. G.P. Jessup During Stone & Webster Conowingo Site Inspection

Chapter Twelve

National Monument

Conowingo's reign as hydropower's poster child was short-lived. As the nation abruptly transitioned from the Roaring '20s to the Great Depression of the 1930s, the halcyon days of hydropower ended. Hydroelectric development by investor-owned utilities essentially halted for the entire decade leading to World War II. Some projects already underway as markets crashed in 1929 were completed, but numerous planned projects were abandoned. Not only was financing impossible under prevailing market conditions, but demand for electricity nationwide also reversed trend, declining nearly 17 percent between 1929 and 1932. Perhaps more importantly, public sentiment turned against monopolies and utility trusts, as we saw with the dramatic fall of Samuel Insull.

The landslide election of Franklin D. Roosevelt as president in November 1932 set the stage for large-scale, multipurpose, federally funded hydropower development. With a national unemployment rate of nearly 25 percent when Roosevelt took office, job creation was critical. He immediately focused on hydroelectric projects as a national imperative for creating large numbers of construction jobs, stimulating economic development, and crushing "evil" utility trusts. The federally funded Big Dam Era had arrived.

The most widely publicized hydropower project of the 1930s was Hoover Dam, and it remains the country's most widely recognized hydroelectric facility. Commonly thought of as part of Roosevelt's New Deal, the project actually had Republican roots. This massive dam spanning the Colorado River between Nevada and Arizona was formally authorized prior to the onset of the Great Depression by the Boulder Canyon Project Act, passed by a Republican Congress, and signed into law in December 1928 by Republican President Calvin Coolidge. Construction began in 1931 during the Administration of Coolidge's successor, Republican Herbert Hoover. Despite his belief that excessive federal intervention in the nation's economic crisis would threaten capitalism and individualism, Hoover long had seen the need for the Hoover Dam project and knew the importance of the jobs its construction provided. He personally had been involved in the evolution of the Boulder Canyon Project since his days as Commerce Secretary under Presidents Harding and Coolidge. Roosevelt, however, adopted the project as his own immediately upon becoming president and wrote Hoover out of its history. Roosevelt's Interior Secretary even changed the dam's name from Hoover Dam to Boulder Dam. (By an Act of Congress in 1947, the name was changed back to Hoover Dam.)

The Colorado River drains one of the driest regions in the country, where water is a limited and valuable resource. Since the 1890s, the river has spurred political battles and efforts by promoters, developers, and citizens' groups to control it and use its water for their particular purposes. The river flows through seven states, each coveting it as a source of economic benefit. It then dips into Mexico before finally entering the Gulf of California. Interestingly, the Colorado borders California for a relatively short distance, and very little water flows into the river from California—yet water projects in southern California have demanded the greatest use of the river's water.

All the initiatives leading to the construction of Hoover Dam related to either the agricultural development of the Imperial Valley or the municipal growth of the Los Angeles area. In 1901, the California Development Company had cut an irrigation channel from the Colorado River, just north of the Mexican border, into a canal passing partly through Mexican territory and into the Imperial Valley. So began the transformation of the desert into rich agricultural land.

By 1905, the Imperial Valley had nearly 15,000 residents, seven towns, nearly 800 miles of water distribution channels, and 120,000 cultivated acres. A Southern Pacific rail line traversed the Valley, providing regular freight and passenger service.[1] But all was not rosy thanks to the wild nature of the Colorado and the precarious financial condition of CDC.

First, the river annually deposited an average of about 130,000 acre-feet of sediment into its delta, enough to bury the entire Imperial Valley under 6 inches of sediment.[2] Despite the fact that this silt had, over eons, enriched the Valley's soil, the heavy loads created a constant battle to keep the irrigation channel and canal

---

1 Michael Hiltzik, *Colossus: The Turbulent, Thrilling Saga of the Building of Hoover Dam*, New York: Free Press, 2011, p. 33.

2 *Ibid.*, p. 36.

open. CDC was too strapped financially to do this effectively. In September 1904, CDC cut a new intake 4 miles south of the international border with Mexico to carry water from the river to the irrigation canal, bypassing its heavily silted initial section. This was done before formal Mexican government approval had been obtained, creating a potential international incident. Furthermore, no head gate was constructed. Without a head gate, flow into the canal could not be cut off in the future if necessary.

Early in 1905, CDC, in desperate need of funding, applied to the Southern Pacific Railroad for a loan and was rebuffed. In an audacious move, CDC then directly approached the railroad's president, Edward H. Harriman, for financial assistance.[3] He agreed, and, in June 1905, the Southern Pacific loaned CDC $200,000 (roughly $6 million in today's dollars). Under terms of the loan agreement, the railroad had the right to name three of CDC's seven directors, one of whom was to be president, and a majority of the company's stock was to be placed in escrow as collateral.

Harriman had been told that river action and seasonal flooding had grown the Mexican intake to the point that water had overflowed the banks of the main canal and had begun flowing into the lowest basin of the Valley. He soon would learn that the situation was much more serious than he had been led to believe. When the newly appointed president of CDC, a distinguished civil engineer and one of Harriman's most trusted employees, went to the Imperial Valley, he was alarmed to see that the Mexican intake actually had become a crevasse 125–150 feet wide.

---

[3] Harriman was an ultrawealthy railroad tycoon. He then was president of the Southern Pacific and the Union Pacific and controlled the Illinois Central and other railroads. His son, Averell Harriman, would become the governor of New York and a widely respected American statesman.

The breach continued to grow despite control efforts. By the end of November, the crevasse had widened to 600 feet. Almost the entire river poured through it into the Valley, forming the Salton Sea in the Valley's lowest basin. The Salton Sea eventually would submerge 500 square miles of land to a maximum depth of 78 feet.

As waves of floods continued, Southern Pacific devoted enormous resources to blocking and redirecting the raging torrent. This turned out to be a gargantuan task. More than a thousand men worked on the job at its peak. An initial closure was achieved on November 4, 1906, but the river broke through after a sudden flood a month later. The river finally was successfully diverted back to its original course in February 1907.

While Southern Pacific was battling to control the river's breach after the initial closure had reopened in December 1906, Harriman had appealed directly to President Theodore Roosevelt for the federal government to take over the task. Harriman's request led to some unexpected consequences.

Harriman and Roosevelt knew each other well, dating back to the 1890s when Harriman was a leading New York Republican Party financial supporter and Roosevelt was its ascending candidate. Harriman was a New York delegate to the 1904 Republican National Convention, raised and contributed money for Roosevelt's campaign that fall at Roosevelt's request, advised him on policy matters, attended his daughter's wedding early in 1906, and was on his Christmas card list. Yet it is remarkable that their interaction regarding the Imperial Valley occurred.

In October 1906, Roosevelt turned on Harriman, denouncing him as an "undesirable" citizen, a "wealthy corruptionist," and an "enemy of the Republic," and instigating a very public Interstate Commerce Commission investigation into his past business activities.[4] Roosevelt's Square Deal domestic program included a promise to battle large industrial combinations, or trusts, that threatened to restrain trade. He apparently had decided to vilify Harriman as a dangerous railroad monopolist. No violation of law was found by the ICC, but Harriman's reputation was sullied.

Roosevelt believed that private concerns, specifically CDC and Southern Pacific since it now in effect controlled CDC, had created the problem and that they should fix it. Roosevelt told Harriman he would ask Congress to reimburse the railroad for closing the break, but then, in a January 1907 message to Congress, he stated that:

> A large amount of money (taken by the California Development Company) which might have been used in needed works was expended in advertising and in propounding the enterprise. The claims were not only extravagant, but in many cases it appears that willful misrepresentation was made ... the money thus obtained from settlers was not used in permanent development, but apparently disappeared either in profits to the principal promoters or in the numerous subsidiary companies ... At the present moment there appears to be only one agency equal to the task of controlling the river, namely the Southern Pacific Company, with its transportation facilities, its equipment, and control of the California Development Company and subsidiary companies. The need of railroad facilities and equipment and the international complications are such that the officers of the United States, even with unlimited funds, could not carry on the work with the celerity re-

---

4 George Kennan, *E.H. Harriman, A Biography, Volume 2*, Cambridge, MA: The Riverside Press, 1922, p. 201.

> quired ... The question as to what sum, if any, should be paid to the Southern Pacific Company for work done since the (closure of the) break of Nov. 4, 1906, is one for future consideration; for work done prior to that date no claim can be admitted ... It is not believed that a free gift of this money should be made, as by its investment the stability of property of great value will be secured and the increase in land values throughout the Imperial Valley will be sufficient to justify the provision that this money should be returned to the Government.[5]

Roosevelt concluded his January 1907 message to Congress by proposing that the Reclamation Service be authorized to "enter upon a broad, comprehensive scheme of development for all the irrigable land upon the Colorado River with needed storage at the head waters so that none of the water of this great river which can be put to beneficial use will be allowed to go to waste."[6]

Before this could happen, Reclamation, Congress, the seven states through which the Colorado flows, Mexico, and an assortment of other interested parties would spend twenty years grappling with the details.

Southern Pacific expended more than $3.1 million (about $86 million in today's dollars) in its efforts to control the Colorado, half of which was spent after the initial redirection of the river in November 1906. In 1930, twenty-three years after Roosevelt's reimbursement assurances to Harriman, Southern Pacific recovered $1 million from the federal government.

---

5 Text of Roosevelt's message to Congress reprinted in *The New York Times*, January 13, 1907, pp. 1–2.

6 *Ibid.*, p. 2.

That was more than CDC got. Lawsuits and flood damage claims dealt CDC its final blow. The company was placed in receivership in 1909. Most of its assets already had been absorbed by Southern Pacific by that point. Then, in 1911, Valley voters formed the Imperial Irrigation District to operate the regional water supply system. Upon purchasing the parts of the system held by Southern Pacific five years later, this public entity assumed sole responsibility for delivering water to the Valley's residents.

It was inevitable that the Colorado River would bring together the interests of Imperial Valley settlers and the broader interests of the fledgling Reclamation Service. After Roosevelt rebuffed Harriman's request for assistance in the Imperial Valley, there were repeated requests for Reclamation to step in. Valley residents worried about the impact of future uncontrolled flooding as well as the effects of dry spells. The obvious solution was an upstream dam and reservoir.

Furthermore, the passage of the Valley's irrigation canal through Mexican territory for about 50 miles led to major issues. Dealing with the Mexican government was difficult. Political unrest in Mexico in 1911 and again in 1914 threatened to cut off the Valley's water supply. CDC's agreement giving Mexico the right to half of the water flowing through the canal rankled Valley farmers when water was rationed during a 1916 drought. People in the Valley clamored for an all-American canal and asked Reclamation to build one.

Even before the organization of the Reclamation Service in 1902, the US Geological Survey had started investigating the lower Colorado River. Upon the formation of the Reclamation Service, Arthur Powell Davis, then its assistant chief engineer, initiated studies of how to develop the resources of the river.

His uncle was John Wesley Powell, who famously led the initial explorations of the upper Colorado River and the Grand Canyon, served as the second director of the Geological Survey, and became a pioneering advocate of irrigation in the West. Davis had accompanied his uncle on various explorations. He continued to pursue his dream of developing the Colorado River in accordance with his uncle's principles when he was named Reclamation's director in 1914. Circumstances converged to create an opening for action after the end of World War I.

In 1919, Reclamation was completing major projects and had time and resources available to pursue Colorado Basin initiatives. Veterans' groups were advocating for expanded irrigated homestead acreage in the West for demobilized soldiers. Los Angeles was growing rapidly and needed water and power to support that growth. Also, that year, the Imperial Irrigation District turned to Washington, DC, seeking a federal guarantee on a $30 million bond to build an all-American canal. As a result of the Irrigation District's political activism and strong encouragement from Davis, who linked the all-American canal issue with the problem of flood control and water storage in the entire Colorado Basin, Congress passed the Kincaid Act in 1920 authorizing Reclamation to develop preliminary plans for a comprehensive Colorado Basin flood control program.

Results of these plans were presented two years later in the Fall/Davis Report.[7] Arthur Powell Davis opened the report with the assertion that reclamation in the Colorado Basin was a federal obligation. The report concluded that an all-American canal and a huge reservoir, and a consequently high dam, would

---

7 The report was principally authored by Davis, under the auspices of Secretary of the Interior Albert B. Fall. Fall soon would become infamous for his involvement in the Teapot Dome scandal.

be necessary to ensure permanent flood control for the Imperial Valley. In accordance with past Reclamation practice, Davis proposed that hydroelectric power production be made part of the project so that construction costs could be repaid over time by the federal government's sale of the resulting hydropower. The hydropower requirement led him to limit potential dam sites to no more than 350 miles from the large and growing southern-California market. This was the approximate practical limit for transmission of electricity at the time. In addition, the dam site would have to be far enough downstream that its impounded reservoir would not back up into the Grand Canyon.

The various requirements resulted in considering dam sites in the Boulder Canyon region, about 300 miles northeast from Los Angeles. In addition to sites in Boulder Canyon itself, others were considered as a backup in the Black Canyon about 20 miles farther downstream. Both had deep, narrow gorges topographically ideal for a high dam. Being farther downstream, Black Canyon offered additional water storage capacity, head height, and hydropower generation capability. It also was 20 miles closer to the nearest town, Las Vegas, which would make construction logistics easier. Although the project would continue to be known as the Boulder Canyon Project, the dam site finally selected was a Black Canyon site.

Leaders of each of the seven states through which the Colorado River passes were highly suspicious of others' possible claims on the river's water. Although California contributed the least amount of water to the river, it by far consumed the most. Before the Boulder Canyon Project ever could proceed, the seven affected states had to agree on how to allocate the river's water among themselves.

Fearing endless litigation, they formed the Colorado River Commission, with a representative of each state and chaired by Herbert Hoover, then President Harding's Commerce Secretary, to try to agree upon a compact of the states. Article I of the Constitution allows states to negotiate compacts among themselves subject to approval of Congress, but this was the first time more than two states had attempted to do so. Distrust of and enmity toward California threatened to make any agreement impossible. Initial talks ended in stalemate. Then, in June 1922, the Supreme Court handed down the *Wyoming v. Colorado* decision, ruling that the law of prior appropriation of water applied regardless of state lines.[8]

As a result, fast-growing California could establish priority use of Colorado River water to the extreme disadvantage of slower growing states in the upper river basin. Fearing this would happen, the seven states quickly reached agreement, permanently allocating a share of the river's water to the states in the upper river basin. The compact was signed that November. It still required official ratification by each state and formal Congressional approval. Six of the states proceeded to ratify the Colorado

---

[8] Water rights are complicated and the source of seemingly constant litigation. Each state has its own set of laws regarding these rights. There are two principal methods of apportioning the use of water—riparian rights (land based), and prior appropriation rights (use based). Under riparian doctrine, water rights belong to landowners whose land physically touches a water source. Under the doctrine of prior appropriation, water is publicly owned, and the first person to divert it for beneficial use has priority over those who come later. Water rights under the prior appropriation system can be sold separately from the land. Riparian apportionment is primarily used in the eastern part of the United States where surface waters are more plentiful. In the western parts of the country, the prior appropriation approach prevails.

River Compact. Arizona, however, ever wary of California, held out and withheld its ratification.

Although Davis had incorporated hydropower in his Boulder Canyon Project plans in order to make the project financially viable, hydropower also was a rallying point for powerful opposition. These were the days of the battle between investor-owned and publicly owned utilities. The Boulder Canyon Project and Ford's Muscle Shoals proposal were being considered simultaneously in Washington. Opponents of federally owned hydroelectric facilities advocated a low dam specifically for flood control as opposed to a larger-scale high storage dam incorporating hydropower. Among the opponents of a high dam was Herbert Hoover.

Opponents argued that a high storage dam would be an extravagant expenditure since there would be insufficient demand for the electricity it would generate. This argument was dispelled early in 1924 when William Mulholland, the infamous Chief of the Los Angeles Bureau of Water Works and Supply, appeared before a Congressional committee and brusquely announced, "I am here in the interest of a domestic water supply for the city of Los Angeles."[9] He explained that the Los Angeles region had been growing rapidly (the population having grown from 670,000 people in 1910 to nearly 1.1 million in 1920) and that he expected rapid growth to continue. He went on to say that because growth depended upon a reliable water supply, Los Angeles intended to construct a regional Colorado River aqueduct, that his plans would require enormous quantities of electricity, and that he proposed to buy all of it from the Boulder Canyon Project.

---

[9] House Committee on Irrigation and Reclamation, *Hearings on Protection and Development of Lower Colorado River Basin, H.R. 2903*, 68 Cong., 1 sess., 1924, p. 97.

William Mulholland became well known when he supervised construction of the 233-mile-long Los Angeles Aqueduct to deliver water from the Owens Valley to the San Fernando Valley in the Los Angeles area. The aqueduct became operational in 1913 and prompted what became known as the California Water Wars.

A man who was talented, imperious, devious, and sometimes underhanded in achieving his goals, Mulholland also was very popular with Angelenos. At the height of his career, he was considered a mayoral candidate.

But his career ended abruptly in 1928 when the St. Francis Dam, located on the Aqueduct just north of the San Fernando Valley, collapsed about twelve hours after Mulholland personally had inspected it. The flood wave swept away thousands of acres of fertile land and emptied bodies and debris into the Pacific Ocean 54 miles away. At least 450 people died in what has been called one of California's worst disasters.

The classic 1974 movie *Chinatown* is but one telling of those events.

Los Angeles subsequently formed the Metropolitan Water District of Southern California to be able to contract for Colorado River water and power.

Samuel Insull and the investor-owned utility industry lobbied hard against the Boulder Canyon Project. The National Electric Light Association alone spent $400,000 opposing the dam. The industry's opposition was understandable. Before Hoover Dam was constructed, Reclamation had built hydroelectric facilities with a total generating capacity of 90,000 horsepower.[10] Privately financed capacity in the West then was more than 3.5 million horsepower. A federally financed Hoover Dam with ultimate generating capacity of more than 1 million horsepower clearly threatened private power's dominance in the region.[11] [12]

Another leading opponent of the Boulder Canyon Project was Hoover's good friend Harry Chandler, the king of southern California business and politics, who owned vast acreage in Mexico (840,000 acres) irrigated by the Imperial Valley canal. An all-American canal imperiled that property. Chandler used his influence on California banks to stymie Imperial Irrigation District efforts to finance an all-American canal.

After years of intense infighting, high-dam advocates prevailed, and President Coolidge signed the Boulder Canyon Project Act in December 1928. Harry Chandler, to everyone's surprise, reversed course on the eve of passage and thenceforth strongly supported the Boulder Canyon Project. It is believed that he had come to recognize that the benefits to his kingdom in southern California would far outweigh any negative impact on his Mexican land holdings. And, in fact, Mexico expropriated hundreds of thousands of acres owned by his syndicate in the mid-1930s.[13]

---

[10] Hay, *Hydroelectric Development in the United States,* p. 128.

[11] David Billington, Donald Jackson, and Martin Melosi, *The History of Large Federal Dams: Planning, Design, and Construction in the Era of Big Dams,* Denver, CO: US Department of the Interior, 2005, p. 165.

[12] It is ironic that in 1902 the Edison Electric Company of Los Angeles determined that there was insufficient potential for profitable hydroelectric power generation on the lower Colorado River due to the then-prevailing 80-mile limit for practical high-voltage transmission and consequently relinquished its options on Colorado River dam sites. This included its option on the Hoover Dam site.

[13] As publisher of the *Los Angeles Times,* community builder, and large-scale

The Boulder Canyon Project Act authorized construction of a dam, in either Boulder or Black canyon, adequate to create a storage reservoir with capacity of at least 20 million acre-feet of water, construction of a hydropower plant at the dam to generate electricity as a means to pay for the project, and construction of the All-American Canal. The Act mandated that construction of the dam and power plant could not start until necessary power contracts were signed for dam cost repayment within fifty years. Costs for the canal were to be repaid over time by owners of its irrigated land.

Recognizing that Arizona was balking at signing the Colorado River Compact, the Act required ratification by six of the seven states party to the compact, one of which had to be California. Congress had modified the original compact to specify the division of lower basin waters and to limit California's allocation.

The Act left the decision to the Interior Secretary of who would install and operate the power plant. Secretary Ray Lyman Wilbur later authorized the federal government (i.e., Reclamation) to build the power plant and required that both a private company (Southern California Edison) and a public entity (Los Angeles Department of Water and Power) operate the facilities. These operators could sell the generated electricity to themselves and other public and private enterprises for resale.

The Act appropriated $165 million. The modified compact was ratified in March 1929 (Arizona, as predicted, withheld its

---

real estate investor, Chandler was motivated to bring water and power to the Los Angeles area. Syndicates he formed owned and developed much of the San Fernando Valley, the Hollywood Hills (he erected the landmark huge, white HOLLYWOODLAND sign), Mulholland Drive, Dana Point, and the Tejon Ranch. At one point, he was the largest private landowner in the country.

approval—until 1944). Contracts for $347 million in power sales were signed in April 1930 after a complicated and contentious process, and Reclamation was authorized to proceed with construction in July of that year. Given the nation's economic plight and burgeoning unemployment, President Hoover ordered Reclamation to expedite preparation of plans and specifications so that the construction contract could be awarded as soon as possible.

Reclamation released bid documents to 107 firms in January 1931. Five bids were submitted on March 4, of which only three were found to be qualified. The low bidder was Six Companies Inc., a joint venture of the Utah Construction Company of Ogden, Utah; Henry J. Kaiser & W. A. Bechtel Company of Oakland, California; McDonald & Kahn Ltd., of Los Angeles; Morrison–Knudsen Company of Boise, Idaho; J. F. Shea Company of Portland, Oregon; and the Pacific Bridge Company of Portland, Oregon. Its $48,890,995 bid somehow, amazingly, was only $24,741 over Reclamation's confidential internal project estimate. The next higher bid, $5 million above Six Companies' bid, was from a group led by the Arundel Corporation of Baltimore. Arundel had worked with Stone & Webster on the construction of Conowingo Dam just three years earlier. The construction contract awarded to Six Companies on March 11 was at the time the largest ever issued by the federal government.

The trump card behind Six Companies was Frank Crowe, whom it had named general superintendent for the Hoover construction project. Crowe was considered the premier dam builder in the West. At the end of his junior year as a civil engineering student at the University of Maine, he spent Summer 1904 working on a Reclamation survey crew along the Yellowstone River. Upon graduation a year later, he began a twenty-year career with Reclamation that briefly was interrupted by two stints with private construction companies that had close ties with Reclamation. In 1925, chafing mightily at a desk job as head of all Reclamation construction and amidst changes occurring at Reclamation upon the naming of a new director, Crowe left the government to join dam constructor Morrison–Knudson. By the time Six Companies submitted its bid for Hoover, Crowe had been involved during his career in the construction of fourteen dams, as superintendent of construction for five of them.

Crowe became known as a gifted deployer of men, material, and equipment and as a decisive, supremely confident and daring problem solver. His relentless drive to move at breakneck speed earned him the nickname "Hurry Up Crowe." He was imposing on the worksite— a lanky man over 6 feet tall with a stern, impersonal demeanor and who wore a large Stetson hat and his trademark pressed white shirt. Crowe had long had his eyes set on building Hoover Dam. As he said in an interview in *Fortune* magazine in 1943, "I was wild to build this dam."[14] Given his close ties to and long experience with Reclamation, there was good reason for Reclamation to want to work with him on the project. Because of this, it likely was no coincidence that the low bid from Six Companies was not far off from Reclamation's internal project cost estimate.

Reclamation's yen to work with Crowe in no way meant that everything on the project would go easily or exactly as planned. The Hoover Dam project presented major challenges. The world's tallest dam had to be shoe-horned into a narrow, nearly vertical canyon situated in an area where summer temperatures regularly reached 120 degrees and created brutal working condi-

---

[14] "The Earth Movers I," *Fortune*, Vol. 28, No. 2, August 1943, p. 103.

tions. Since Hoover Dam at 726 feet would be over twice as tall as any existing dam, its designers and builders were venturing into uncharted territory.[15]

This uncharted territory required workers to perform with extraordinary effort. To groom the tall canyon walls abutting the dam, for example, high scalers suspended from long ropes brandished 44-pound jackhammers to clear the walls of debris. One spectacular feat performed by high scalers Oliver Cowan and Arnold Parks highlights the dam site's treacherous work environment and the amazing skill of high scalers:

> One afternoon a Bureau of Reclamation engineer, Burl Rutledge, was trying to inspect a portion of the cliff that had just been cleared of protruding rock when he leaned out too far, slipped, and began to roll and tumble down the precipitous slope toward the river far below. Cowan, who was working some twenty-five feet beneath the spot where Rutledge had been standing, heard the engineer's muffled exclamation, looked up, and saw him begin his fall. Without hesitation, he pushed out from the wall, propelled himself horizontally through the air, swung back in to the cliff face, and snagged the falling man's leg. Parks, who had seen the accident unfolding, swung over seconds later and pinned Rutledge's upper body to the canyon wall. The two scalers then held the stunned engineer in place until a line could be rigged to haul him back to safety.[16]

---

15 As the dam's design was undertaken, Reclamation's Arrowrock Dam in Idaho, completed in 1915, had been the world's tallest dam at 348 feet. Crowe had been integral to Arrowhead's construction. As Hoover was being constructed, Reclamation's 417-foot-tall Owyhee Dam in Oregon became next tallest upon its completion in 1932.

16 Joseph E. Stevens, *Hoover Dam: An American Adventure*, Norman, OK: University of Oklahoma Press, 1988, pp. 106–107.

The Colorado River's variable flow and sudden flooding added further complexity. To be able to divert the river's flow around the dam's construction, four tunnels 56 feet in diameter were driven through the solid rock canyon walls, two on each side of the river. Their combined length was over 3 miles. The diversion tunnels were bored from both ends simultaneously to accelerate the process. Eight unique "drilling jumbos" were assembled and mounted on the back of gasoline-powered, 10-ton trucks. The trucks backed up to the working rock face, thirty holes were drilled at once, and the holes were filled with dynamite to be exploded in unison.

In addition, the huge amount of concrete to be poured to form the dam and powerhouse—about 4.4 million cubic yards—and the heights from which it was to be poured presented mammoth challenges. As noted in Chapter Five, concrete produces heat as it cures, which can lead to cracking. Cure time increases with the amount of concrete. Bureau of Reclamation engineers calculated that if the dam had been formed in a single, continuous pour, its internal temperature would have risen 40 degrees while it was hardening, it would have taken 125 years to cool, and the thermal stresses would have fractured it severely.[17] Instead, the dam was formed of 5-foot-high interlocking blocks of concrete, in an extension of a technique first used in 1889 to construct the San Mateo Canyon Dam near San Francisco (see Chapter Five). An intricate network of water-cooling lines was embedded in each block. These passageways were filled with grout after the concrete set. To ensure structural integrity of the dam, the concrete itself had to be a uniform product produced to exacting specifications. An elaborate aerial cableway system crossing above the

[17] *Ibid.*, p. 193.

canyon was used to quickly deliver buckets containing 8 cubic yards of freshly mixed concrete with pinpoint accuracy to fill block forms atop the rising dam. (Crowe had developed the first version of this delivery system while building Arrowrock Dam nearly twenty years earlier.)

Building Hoover Dam was exceedingly complex. As Crowe put it, "We had 5,000 men in a 4,000-foot canyon. The problem, which was a problem in materials flow, was to set up the right sequence of jobs so (the workers) wouldn't kill each other off."[18]

Although Crowe was a master organizer, his relentless focus on speed exacerbated safety issues. At least two hundred men died during the dam's construction, including a number from heat prostration and carbon monoxide poisoning while the diversion tunnels were being bored. That same focus on speed, however, made Crowe a wealthy man. The Six Companies construction contract included incentives for early completion. Crowe brought the project in more than two years ahead of schedule. On top of his $18,000/year salary, his performance bonus is estimated to have been as much as $450,000 (about $8.5 million in today's dollars).[19]

On February 1, 1935, the last diversion tunnel gate was closed and the dam's reservoir, Lake Mead,[20] began filling. On March 1, 1936, the federal government accepted the dam and powerhouse, marking the end of the Six Companies construction contract. Construction might have ended two years ahead of schedule, but improvements continued for years. Installation of the seventeen main turbine-generators in Hoover's power-

18 *Ibid.*, p. 195.

19 *Ibid.*, p. 252.

20 Named in honor of Dr. Elwood Mead, Commissioner of the Bureau of Reclamation, 1924–36.

houses extended over a period of years as demand increased. The first turbine-generator unit went into operation in October 1936. By the end of 1939, nine units were operational, with an installed capacity of 704 MW, making Hoover Dam the largest hydroelectric powerplant in the world (a distinction held until surpassed by Grand Coulee in 1947). The final unit did not come online until 1961, bringing capacity to 1,335 MW.

In 1938, about 275 miles south of Hoover Dam near Yuma, Arizona, Imperial Dam was completed to divert water into the All-American Canal. The canal, then under construction, became operational in 1940, fulfilling the long-time dream of Imperial Valley farmers.

Parker Dam, constructed between 1934 and 1938 by Reclamation 155 miles downstream from Hoover Dam, began feeding water to the Los Angeles Metropolitan Water District's Colorado River Aqueduct in 1941. The Aqueduct's pumping stations are driven by power transmitted from Hoover Dam.[21]

The Hoover Dam saga continues today. In 2018, the City of Los Angeles announced it was studying a major Hoover Dam pumped energy storage project as a key component in achieving California's green energy goals. Unlike typical pumped hydro projects, the project would use an external source of power to pump water back up to Lake Mead, and it would be independent of Hoover Dam's physical plant. During times of excess carbon-free renewable energy generation, pumps powered solely by wind and solar energy would propel water 20 miles through

---

[21] After completion of Hoover Dam, Frank Crowe supervised the construction of Parker Dam. His last dam was Shasta Dam, a 602-foot-high, curved gravity dam built on the Sacramento River near Redding, California. He retired in 1945 to his 13,000-acre cattle ranch nearby. He died of a heart attack a year later at the age of 63.

a new pipeline to Lake Mead to be released through the dam's existing turbines when electricity demand was high. Targeted completion for the $3 billion-plus project was 2028–30, but the proposal faced significant approval hurdles.

Hoover Dam is considered a national treasure, heralded as a monument to American resourcefulness and as a model for harnessing forces of nature through mammoth technological solutions. Hoover was the first of what came to be characterized as large, multipurpose dams. This national monument is a source of national pride (and annually welcomes more than 7 million tourists). It revolutionized the way the federal government participated in water control projects. It symbolizes man's mastery over nature for the common good and the nation's ability thereby to find greatness. As Roosevelt put it when he dedicated Boulder (Hoover) Dam in September 1935:

> This morning I came, I saw, and I was conquered, as everyone would be who sees for the first time this great feat of mankind... We are here to celebrate the completion of the greatest dam in the world, rising 726 feet above the bedrock of the river and altering the geography of a whole region... We know that, as an unregulated river, the Colorado added little of value to the region this dam serves. When in flood, the river was a threatening torrent. In the dry months of the year, it shrank to a trickling stream. For a generation the people of the Imperial Valley had lived in the shadow of disaster from this river which provided their livelihood, and which is the foundation of their hopes for themselves and their children. Every spring they awaited with dread the coming of a flood, and at the end of nearly every summer, they feared a shortage of water would destroy their crops... This great Boulder Dam warrants universal approval because it will prevent floods and

flood damage, because it will irrigate thousands of acres of tillable land, and because it will generate electricity to turn the wheels of many factories and illuminate countless homes... The mighty waters of the Colorado were running unused to the sea. Today we translate them into a great national possession... This is an engineering victory of the first order—another great achievement of American resourcefulness, American skill and determination.[22]

[22] Excerpted from Franklin D. Roosevelt, *Address at the dedication of Boulder Dam*, September 30, 1935; online by Gerhard Peters and John T. Woolley, The American Presidency Project https://www.presidency.ucsb.edu/documents/address-the-dedication-boulder-dam

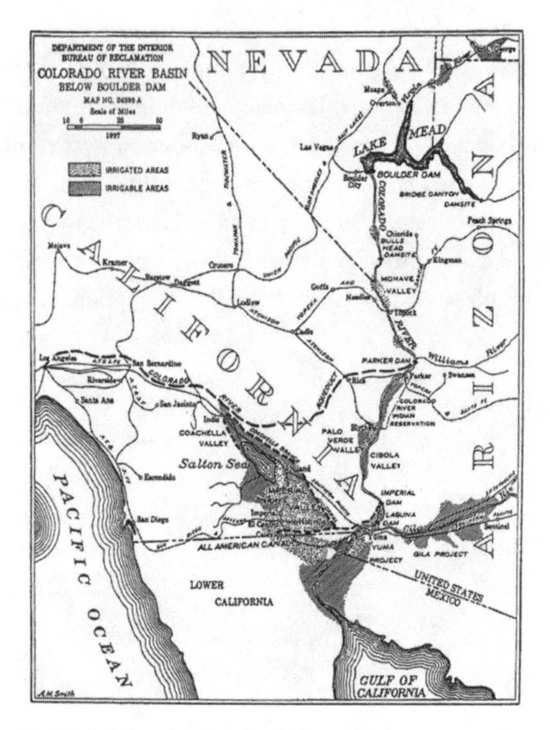

Fig. 12.1. Colorado River Basin Below Hoover Dam in 1937

Fig. 12.2. Colorado River Flowing Unchecked Over 28 Foot Falls Toward Salton Sea

Fig. 12.3. E.H. Harriman, Southern Pacific Railroad President

Fig. 12.4. Arthur P. Davis, Reclamation Director 1914–23

Fig. 12.5. Harry Chandler (left) and William Mulholland
The Los Angeles Water and Power Duo

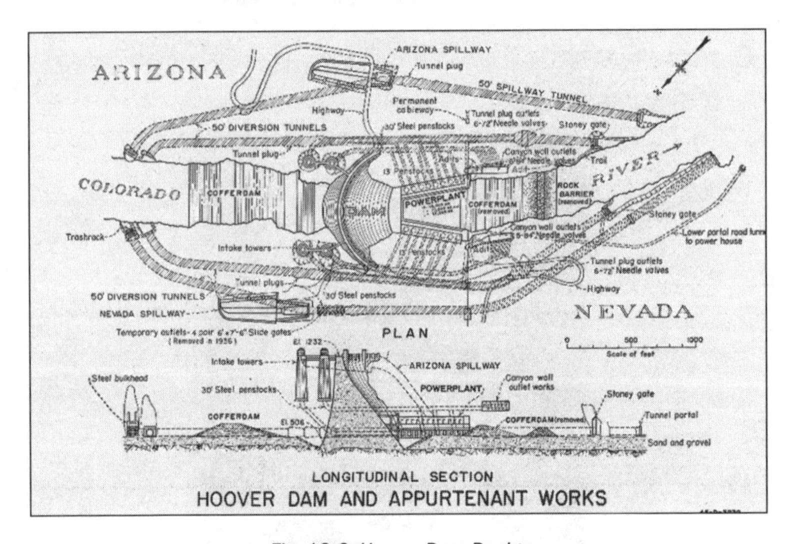

Fig. 12.6. Hoover Dam Design

Fig. 12.7. Frank Crowe at Hoover Dam Site

Fig. 12.8. High Scalers During Hoover Dam Construction

Fig. 12.9. Hoover Dam Diversion Tunnel

Fig. 12.10. Hoover Dam Site Before Construction

Fig. 12.11. Hoover Dam upon Completion

Chapter Thirteen

Hail Columbia

Hoover Dam paved the way for two other significant projects on the Columbia River: Bonneville and Grand Coulee. In a campaign speech in Portland shortly before winning the 1932 presidential election, Franklin D. Roosevelt had proclaimed, "I state, in definite and certain terms, that the next great hydroelectric development to be undertaken by the federal government must be that on the Columbia River."[1] Upon taking office in March 1933, he moved quickly to make that happen. Under the auspices of the hastily passed National Industrial Recovery Act, he authorized the Bureau of Reclamation's Grand Coulee project in north–central Washington in July 1933 and the Corps of Engineers' Bonneville project near Portland in September.

---

[1] Franklin D. Roosevelt policy speech, Portland, Oregon, September 21, 1932. For the complete text of the speech, see http://www.fdrlibrary.marist.edu/_resources/images/msf/msf00530.pdf .

Harnessing the Columbia had been contemplated by many different groups and factions for years—for irrigation in vast areas of north–central Washington, for hydropower, for navigation, and for flood control. The fourth largest river in the United States by volume,[2] the Columbia drops 2,700 feet along its 1,240-mile course. In Washington alone, it drops more than 1,000 feet over about 400 miles. With passage of the 1925 River and Harbor Act, Congress took the first steps toward eventually damming the Columbia and other major American rivers by ordering the Corps and the Federal Power Commission to jointly prepare a cost estimate for a national survey of navigable streams "with a view to the formulation of general plans for the most effective improvement of such streams for the purposes of navigation and the prosecution of such improvement in combination with the most efficient development of the potential water power, the control of floods, and the needs of irrigation."[3] After the Corps submitted a plan for such surveys (the so-called original *308 Report* due to the document number assigned to the report by the House of Representatives), Congress in the 1927 River and Harbor Act authorized the Corps to conduct detailed studies of selected river basins, including the Columbia. The resulting 1,845-page survey report titled *Columbia River and Minor Tributaries (House Document 103)* was submitted to Congress in March 1932. The report recommended a ten-dam comprehensive plan for the Columbia River, with Grand Coulee as the farthest upriver project and Bonneville as the lowermost. The ten dams included one dam already under construction: the Rock Island Dam being built by

---

[2] The top three (in descending order) are the Mississippi River, the St. Lawrence River, and the Ohio River.

[3] River and Harbor Improvements Act of 1925, Public Law No. 68–585, Ch. 467, 43 Stat. 1186, Sec. 3, March 3, 1925.

Stone & Webster/Puget Sound Power and Light.[4] A year later, the report gave newly elected President Roosevelt the justification he wanted to authorize the giant, federally funded Bonneville and Grand Coulee construction projects to put people back to work and help begin a national recovery.

There was a flurry of last-minute maneuvering before the Bonneville and Grand Coulee projects were authorized. The Grand Coulee project had been planned to include a 550-foot-tall dam across the Columbia generating hydropower capable of pumping river water up over 500 feet into the Grand Coulee. This 50-mile-long canyon formed a natural reservoir, from which water would be fed by gravity through canals to irrigate roughly a million acres of dry, potentially fertile land in central Washington's Columbia Basin.

Roosevelt balked when he was told that the cost estimate for the project was $450 million (more than $9 billion in today's dollars), pointing out that the amount was more than the cost of the Panama Canal, which was completed in 1914. He instead supported initially stressing the power aspects of the project via a cheaper, more modest low dam (290 feet tall) that could be added onto later to provide additional power capacity to accomplish the project's irrigation goals. That approach made the project more palatable politically since low-dam project costs could be recouped through hydroelectricity sales. For this he would authorize $63 million for the project to proceed.

Leaders in Oregon, sensing that Roosevelt might support only one dam, mobilized to ensure Bonneville funding. The Bonneville project, with a dam and ship lock to be located on the Columbia near the Cascades Rapids about 40 miles east of Port-

[4] See also Chapters Six and Nine.

land, would allow oceangoing vessels to pass another 48 miles upriver from that point. It also would provide substantial hydropower to support the region's electricity demand. The 1932 *Columbia River and Minor Tributaries* survey report had identified nearby Warrendale as the dam site, but there were concerns over foundation conditions there. Roosevelt made it clear that he would not commit federal funds to the Bonneville project unless he could be guaranteed that a suitable foundation existed.[5]

During Summer 1933, the Corps made more detailed geological surveys, determined that suitable conditions existed at Bonneville, and submitted a favorable report on Bonneville to Interior Secretary Harold Ickes. Ickes opposed the Bonneville project and did not transmit the report to Roosevelt. In September, Oregon Senator Charles McNary,[6] a Republican, and Oregon's 3rd Congressional District Representative, Democrat Charles Martin, bypassed Ickes, met privately with Roosevelt, showed him the favorable Corps report, and argued the case for Bonneville.[7] Roosevelt quickly thereafter approved the project and authorized $20 million in initial funding.

In the short time between assuming office and initiating the Bonneville and Grand Coulee projects, FDR was able to cut through conflicts between Oregon and Washington interests, competitive maneuvers between the Corps and Reclamation, factions within Congress, and opposition from detractors of public/federal waterpower development. His actions clearly

---

5 David Billington and Donald Jackson, *Big Dams of the New Deal Era*, Norman, OK: University of Oklahoma Press, 2006, p. 157.

6 McNary was Senate Minority Leader from 1933–44. In 1940, he was the Republican vice presidential candidate, Wendell Willkie's running mate in his unsuccessful bid to unseat Roosevelt.

7 Billington and Jackson, *Big Dams of the New Dam Era*, p. 170.

established federal authority over the waters of the West. The master politician also sidestepped Congress to start both projects by providing money out of his general funds. The projects began by presidential decree. He left it to Congress to later continue to pay for them or abandon them unfinished.

The Bonneville and Grand Coulee projects began almost simultaneously to great national fanfare publicizing that they would provide sorely needed jobs and would generate plentiful cheap electricity to spur regional economic development (and, in the case of Grand Coulee, later to create a vast area of fertile agricultural land via irrigation). Both projects did create jobs. When Bonneville construction was fully underway, the total work force averaged about three thousand. Peak employment for Grand Coulee construction was estimated to be around eighty-eight hundred workers.

Ongoing hype for Grand Coulee eclipsed that for Bonneville due to Grand Coulee's staggering scope, even though Bonneville was completed five years sooner and was capable of producing impressive amounts of electricity. The generating capacity of Bonneville's original powerhouse was 518 MW, more than twice that of Conowingo. Conowingo had been the second largest hydroelectric plant in the world when completed in 1928.[8]

When Roosevelt authorized the Bonneville project in September 1933, planning had not progressed beyond the preliminary study and investigation stage. To generate jobs as quickly as possible, the Corps divided the project into a large number of discrete elements and rushed to award construction contracts for

[8] Critics continued to argue that there was no market for all the electricity the two dams would be able to produce and that building them thus was a terrible waste of money. It would take World War II to prove the critics wrong.

each element as plans were developed. Columbia Construction Company, a joint venture of participants in the Six Companies venture of Hoover Dam fame, began work on the spillway dam in June 1934.[9] A month later, a separate contract was awarded for building the lock and powerhouse substructure.

The dam's designers took advantage of its location on the Columbia River at the Cascades Rapids. An island there divided the river's flow. A 1,450-foot-long, run-of-the-river concrete spillway dam was constructed, extending from the island to the Washington side of the river. To ensure stability on the weak volcanic andesite foundation rock at the site, large notches, or saw teeth, in the dam's base supplemented the dam's weight in anchoring the dam. The dam was 132 feet wide at its base and 197 feet high above the lowest bedrock. To be able to pass large floods without causing a material rise in head water elevations, the spillway contained eighteen 200-ton steel gates that were 50 feet wide. Of those, twelve were 50 feet tall; six were 60 feet tall. When fully raised, these exceptionally large gates could pass a flood 37 percent larger than ever recorded on the Columbia. The spillway cross section in Figure 13.1 shows its unique saw-tooth base and tall gates.

The powerhouse and ship lock were placed in the channel between the other side of the island and the Oregon riverbank. The 1,027-foot-long powerhouse contained ten generators, placed in operation in 1938–43, to give the ultimate generating capacity of 518 MW. Each generator was attached to a Kaplan adjustable-blade turbine. Kaplan turbines were a recent breakthrough

---

9 Six Company participants Pacific Bridge Co. and J.F. Shea Co. did not join in formation of Columbia Construction Company. J.F. Shea participated in a separate contract awarded in October 1935 for construction of the powerhouse superstructure.

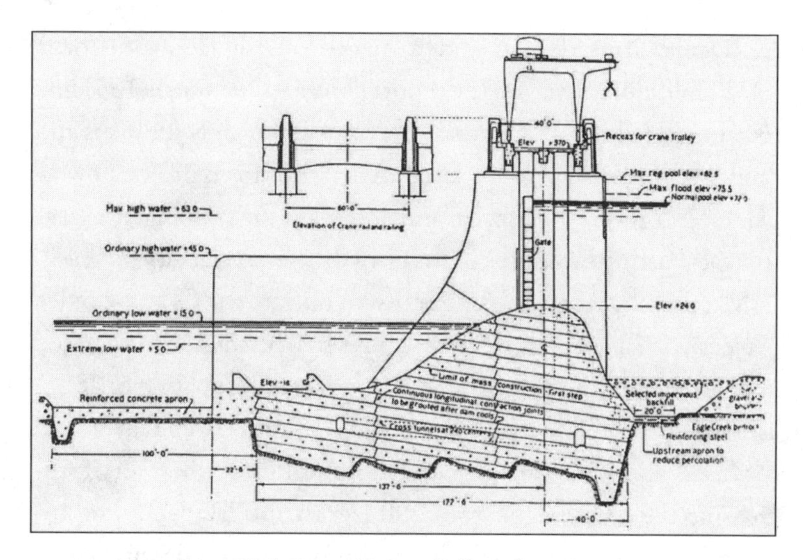

Figure 13.1: Bonneville Dam Spillway Cross Section

in turbine technology. They achieve maximum efficiency under a wide range of load and head (see Chapter Five). The Bonneville turbines were much larger and more powerful than any Kaplan turbines previously installed in the world.[10] In the 1980s, a second powerhouse was added. Today, Bonneville's generating capacity is a walloping 1,227 MW, enough electricity to power nearly a million homes. The 60-foot single-lift navigation lock adjacent to the powerhouse had a chamber 500 feet long and 76 feet wide. Bonneville also incorporated an extensive fish-collection and -passage system.

The Bonneville project was of major significance for the Corps. Until then, the Corps by policy had shied away from hydroelec-

[10] In 1929, the first Kaplan turbine installed in the United States became operational in a facility on the Susquehanna River at York Haven, Pennsylvania (1,360 kW). The first Bonneville turbine unit produced 43,200 kW.

tric facility construction and management. In fact, the Corps' 1932 Columbia River survey report (House Document 103) had recommended that the federal government pay only for locks and dredging (i.e., navigation-related items), not for dams or powerplants. No more: Roosevelt and his New Deal changed everything, thrusting the Corps in to the hydroelectric facility business.

Meanwhile, Roosevelt's directive to initiate the Grand Coulee project with a low-height dam whose height could be increased later to enable the project's original irrigation goals greatly increased the complexity of an already ambitious endeavor and confused the situation. Because no one knew when, or if, increasing the dam height from 290 feet to 550 feet would be approved, construction would have to proceed not knowing when the project might be redirected.

What had been billed as a massive irrigation project had been repositioned as a hydropower project to be financed as a jobs program. This intensified ongoing opposition from forces fighting federal hydropower development and created pressure to quickly create jobs even though design of the low dam was not yet complete. A contract for preliminary excavation was awarded in November 1933 and for building roads and bridges soon thereafter. Reclamation called for bids for the low dam and powerhouses (but no irrigation facilities) on March 3, 1934. Four bids were received on June 18.

Although Six Companies, based upon its experience at Hoover Dam, was considered a shoo-in to win the competition, it was outbid by MWAK Company, a corporation formed by three companies: the Silas Mason Company of New York, Walsh Construction Company of Davenport, Iowa, and Atkinson-Kier Company of San Francisco. MWAK was formally awarded the job on July 13, almost exactly a year after $63 million of Public

Works Administration funds were committed for the low dam Grand Coulee project.

MWAK completed the west cofferdam in April 1935 so that the foundation for the western third of the dam could be poured. In June, Interior Secretary Ickes signed a change order redirecting MWAK to instead build the foundation for the high dam, and the PWA allotted additional funds to carry construction through 1936 for this beginning of the high dam project. Construction of the high dam itself would come later under a separate second contract.

A short time earlier, on April 29, the Supreme Court had dealt Roosevelt a blow that put the Grand Coulee project in serious jeopardy when it ruled that only Congress could authorize construction of dams across navigable waters. The ruling specifically addressed the authorization of Reclamation's Parker Dam project on the Colorado River, but the implications were clear for the twenty-some other federal dams already being built via the PWA. This development greatly intensified the efforts of opponents of Grand Coulee as the Roosevelt Administration rushed to push a new Rivers and Harbors bill through Congress to legitimize the PWA-funded dams.

Among the most vocal opponents was Seattle City Light's J.D. Ross (see Chapter Ten), who continued to argue that Grand Coulee would rob Seattle of a market it deserved, that his Skagit River projects were far cheaper vehicles for providing power to Eastern Washington, and that funds earmarked for Grand Coulee instead should go to Skagit. Ross was known to fight dirty in his attempts to scuttle Grand Coulee. After a bruising battle in Congress in which Ross was a protagonist, Roosevelt signed the new Rivers and Harbors measure authorizing the Grand Coulee project on August 30, 1935.

MWAK began pouring the dam's concrete foundation at the beginning of December 1935 and completed it two years later. By then, Reclamation had prepared specifications for the next phase of construction: completion of the high dam, a west powerhouse, and the foundations for an east powerhouse and irrigation pumping plant. In November, this phase was put out for bid, with bid opening scheduled for December 10. In a stunning move just prior to submittal, Henry Kaiser of Six Companies approached MWAK about joining forces to bid.

The resulting Consolidated Builders Incorporated won the competition and officially was ordered to proceed on March 18, 1936. The first of the dam's primary generators, the world's first 108 MW generator, went online in October 1941. Concrete pouring—all 11 million cubic yards of it—was completed two months later, five days after the Japanese bombed Pearl Harbor on December 7, 1941. The 151-mile-long reservoir behind the dam filled by June 1942. Reclamation assumed possession of the dam in January 1943, almost ten years after Roosevelt initiated the project.

Grand Coulee Dam is immense. The 550-foot-high dam is about a mile long and 500 feet wide at the base. When completed, it was the largest concrete structure in the world; it remains the largest in the United States. As Woodie Guthrie put it in his 1941 song *Roll on Columbia,* "Grand Coulee Dam, the mightiest thing ever built by man."

With the onset of World War II, the focus at Grand Coulee turned to providing hydropower to support wartime aircraft and aluminum industries in the Pacific Northwest as well as to the secret Manhattan Project nuclear facility at Hanford, Washington. By 1947, the output of its power facilities exceeded that of Hoover Dam. The last of the nine generators in the west pow-

A close friend of President Roosevelt, *Henry Kaiser* was mentioned as a possible running mate for FDR in 1944. He used his Washington connections to secure government contracts and loans, amassing over time a huge corporate empire.

Kaiser formed more than a hundred companies. He was a consummate serial entrepreneur and self-promoter who had a real flair for working the halls of government power. His dam-building experience led him to form Permanente Cement, which quickly became the largest cement manufacturer in the world, and Kaiser Steel, soon the largest steel manufacturer on the West Coast. During World War II, Kaiser Shipyards, known for its prowess in building Liberty ships, was responsible for constructing 1,490 vessels.

After the war, Kaiser added automobiles, aviation components, chemicals, healthcare, real estate, and television and radio to his major interests in steel and cement. He is best known for having formed Kaiser Permanente Healthcare.

erhouse went online in October 1947, giving Grand Coulee a rated capacity of 1,002 MW. Wartime power needs set back irrigation for nearly a decade. Pumping for irrigation began in 1951. Currently, 670,000 acres are irrigated.[11]

After the last of the nine generators in the east powerhouse became operational in 1950, total rated capacity rose to 1,974

[11] In comparison, water from the Colorado River passing through the All-American Canal irrigates 500,000 acres in California's Imperial Valley (see Chapter Eleven).

MW. Since then, a third powerhouse and a pumping station have been constructed, and the original generating units have been upgraded. Today, Grand Coulee is the largest hydropower facility in the country, with a total rated capacity of 6,809 MW.

More than 5 million tourists visit Grand Canyon annually to witness the power of Mother Nature in shaping our earth. Many of those tourists also are among the 8 million people who travel to see Grand Coulee, Bonneville, or Hoover dam each year. Every society builds monuments to what it believes is important. Nations always have erected statues of their leaders and warriors. Egyptian tombs enshrined a focus on the afterlife. The Great Wall evidenced concerns about boundaries and security. Grand Coulee, Bonneville, and Hoover dams were twentieth-century national monuments—testaments to American ingenuity and symbols of man's ability to control the forces of nature for the common good.

Yet these dams were more than that. Bonneville and Grand Coulee sprang from the Corps' ten-dam blueprint for the Columbia River. Hoover was just the initial step in developing the waterpower of the Colorado River basin. These multipurpose dams were at the forefront of another phase in the evolution of the power industry: planned federal development to harness the waterways of entire river basins. Among the other watersheds focused on in this era of big federal dams were the Missouri River, California's Central Valley, the Ohio River basin, and the Upper Mississippi. The shining example, however, was the Tennessee Valley. The government's success led one sage to observe that the country's raging rivers had been tamed and turned into series of lakes.

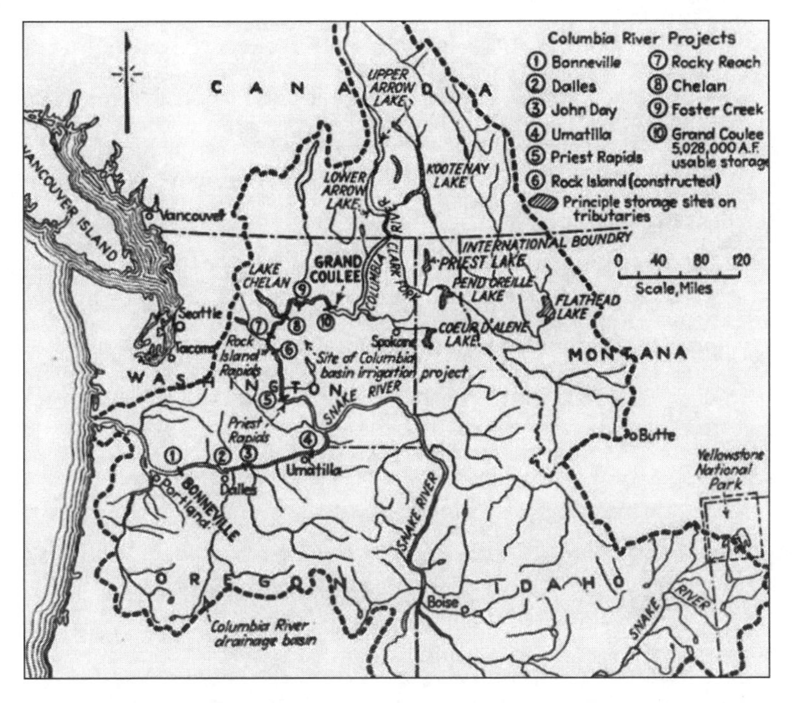

Fig. 13.2. Columbia River Dams Proposed to Congress by the Corps in 1932 (all ten now have been built)

Fig. 13.3. Bonneville Dam Overview circa 1940

Fig. 13.4. Bonneville Dam with Spillways Open

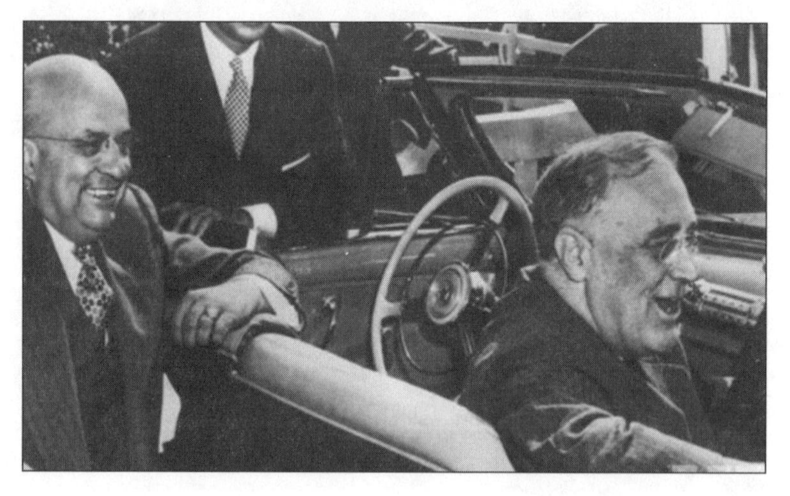

Fig. 13.5. Henry Kaiser and President Roosevelt, 1942

Fig. 13.6. Grand Coulee Excavation Conveyer Belt
(over one mile long)

Fig. 13.7. Grand Coulee Rotor Installation

Fig. 13.8. Grand Coulee Dam circa 1942

Chapter Fourteen

The TVA

Upon leaving home in Tennessee as a teenager to attend school "up North," I was amazed by my new classmates' image of Tennesseans. They envisioned a barefoot man in bib overalls standing in front of an unpainted shack next to a woman in a thread-bare house dress with a toddler on her hip. They clearly were unaware of the renaissance underway in the mid-South, which was stimulated by the creation of the TVA in 1933.

The Tennessee Valley, or Tennessee River Basin, although centered in the state of Tennessee, also covers parts of Kentucky, Virginia, North Carolina, Georgia, Alabama, and Mississippi. The Tennessee River flows westward 652 miles from Knoxville, where it is formed by the confluence of the French Broad and Holston rivers, to its mouth at Paducah, Kentucky, there joining the Ohio River shortly before it runs into the Mississippi River. It is the fifth largest river in the United States measured by flow and drains water from 40,900 square miles in the River Basin.

Since the 1800s, the river had been envisioned as a navigational link between the mid-South and the Mississippi River transportation system. This goal, however, was long impeded by hazardous rapids, islands, reefs, and bars nearly halfway up the river in the Muscle Shoals area that effectively divided the river into two sections. Furthermore, the river was subject to devastating periodic floods. During a flood in 1867, Chattanooga was almost completely submerged when the river crested 28 feet above its banks. The river's periodic flooding also was a major contributor to floods along the Ohio and Mississippi rivers.

Economic conditions in the Tennessee Valley during the Depression days of the early 1930s were abysmal. More than 75 percent of the 2.3 million people in the Valley lived in rural areas, compared with 44 percent nationally. The region was a patchwork quilt of small farmsteads and crossroads communities built around a general store, a church, and a school. Average annual farm income was only $639 (versus a national average of $1,835), with some families surviving on as little as $100 per year. Farmers subsisted on depleted land that had been farmed too hard for too long. Erosion had damaged roughly 85 percent of the Valley's cultivated land. Only 3 percent of Valley farms had running water. Poor sanitary conditions led to high rates of tuberculosis and typhoid. One third of the people in northern Alabama had malaria. At a time when almost 90 percent of the nation's urban population had electricity, only 4 percent of Valley farms did. Private utility companies argued that it was too expensive to run transmission lines to isolated farms and that farmers could not afford electricity anyway.

As we saw in Chapter Eight, Henry Ford's offer to buy the federal government's facilities on the Tennessee River at Muscle Shoals—both Wilson Dam and the nearby nitrate plants—and

Senator George Norris's successful attempts to squelch that purchase drew national attention to the Tennessee Valley, its potential, and its issues. In his relentless crusade to bring public power to the region, Norris twice succeeded in having Congress pass legislation calling for federal ownership and operation of all Muscle Shoals properties as well as for another dam to be built upstream—only to have the bills vetoed by Republican Presidents Coolidge and Hoover. Roosevelt's election changed things.

Roosevelt had made electrical power, especially hydropower, a major campaign theme leading up to the 1932 elections. In a policy speech in Portland on September 21, 1932, he railed against the "Insull monstrosity" and the utility trusts and their financial manipulation, for which "the public paid and paid dearly." He went on to state that, unlike then-President Hoover, he favored giving the federal government the right to operate power businesses where and when essential to protect the people of the United States against inefficient service or exorbitant charges. He painted a picture of four great federal power developments: the St. Lawrence River in the Northeast, Muscle Shoals in the Southeast, the Boulder Dam project in the Southwest, and the Columbia River in the Northwest. Each forever would be "a national yardstick to prevent extortion against the public and to encourage the wider use of that servant of the people—electricity."[1]

Two months after Roosevelt crushed Hoover in the election, Senator Norris accompanied the president-elect to Muscle Shoals so that Roosevelt could "see what the whole Tennessee River project looks like." Only a month after his inauguration, on April 10, 1933, Roosevelt sent a formal request to Congress

---

1 Roosevelt, Portland speech, http://www.fdrlibrary.marist.edu/_resources/images/msf/msf00530.pdf.

to establish the TVA. The scope of his proposal both surprised and delighted Norris:

> TO THE CONGRESS:
>
> The continued idleness of a great national investment in the Tennessee Valley leads me to ask the Congress for legislation necessary to enlist this project in the service of the people.
>
> It is clear that the Muscle Shoals development is but a small part of the potential public usefulness of the entire Tennessee River. Such use, if envisioned in its entirety, transcends mere power development; it enters the wide fields of flood control, soil erosion, afforestation, elimination from agricultural use of marginal lands, and distribution and diversification of industry. In short, this power development of war days leads logically to national planning for a complete river watershed involving many States and the future lives and welfare of millions. It touches and gives life to all forms of human concerns.
>
> I, therefore, suggest to the Congress legislation to create a Tennessee Valley Authority, a corporation clothed with the power of Government but possessed of the flexibility and initiative of a private enterprise. It should be charged with the broadest duty of planning for the proper use, conservation and development of the natural resources of the Tennessee River drainage basin and its adjoining territory for the general social and economic welfare of the Nation. This Authority should also be clothed with the necessary power to carry these plans into effect. Its duty should be the rehabilitation of the Muscle Shoals development and the coordination of it with the wider plan.
>
> Many hard lessons have taught us the human waste that results from lack of planning. Here and there a few

> wise cities and counties have looked ahead and planned. But our Nation has "just grown." It is time to extend planning to a wider field, in this instance comprehending in one great project many States directly concerned with the basin of one of our greatest rivers.
>
> This in a true sense is a return to the spirit and vision of the pioneer. If we are successful here, we can march on, step by step, in a like development of other great natural territorial units within our borders.[2]

Just a month later, on May 18, the president signed legislation establishing the TVA. This public corporation was granted sweeping powers and unprecedented independence. For the first time, the federal government obtained the power to undertake centralized planning for and development of an entire multistate region of the country.Flood control, navigation, and power generation, distribution, and sales were to be means to advance the social and economic well-being of the Valley. Other initiatives included production of nitrates and phosphates for fertilizer and defense needs; malaria control; reforestation; development of mineral, fish, and wildlife resources; land conservation; model housing; rural electrification; educational and social programs; and construction of recreational facilities adjacent to reservoir banks.

The TVA was to have three directors appointed by the president, be headquartered in the Tennessee Valley (first Muscle Shoals and then Knoxville) rather than in Washington, and be able to condemn land and issue government-guaranteed bonds. Private utilities were aghast and furious. Critics blasted the

---

2 For a copy of Roosevelt's message to Congress, see https://www.visitthecapitol.gov/exhibitions/instruments-change/harnessing-nature-generating-jobs.

TVA as blatant socialism. Nonetheless, it would withstand legal challenges all the way through the Supreme Court.

From its beginning, the TVA planned to develop the Tennessee River Basin into a unified river system with a continuous 9-foot-minimum navigable channel along the river's entirety from Knoxville to Paducah. A series of hydroelectric dams and locks and associated reservoirs was necessary to create a channel of the required depth over the river's 652-mile descent.In addition, dams were to be constructed on the main tributaries joining the river from the mountains to the east to produce hydroelectricity and to regulate water flow levels in the main-stem Tennessee River.

By 1945, nine dams owned and operated by the TVA impounded reservoirs that created a stairway of navigable water on the Tennessee River from Knoxville to the mouth of the river at Paducah. Figure 14.1 shows this progression of dams and reservoirs, the mile marker upstream from the mouth of the river where each was located, its elevation in feet above sea level, and the year that dam construction was completed.

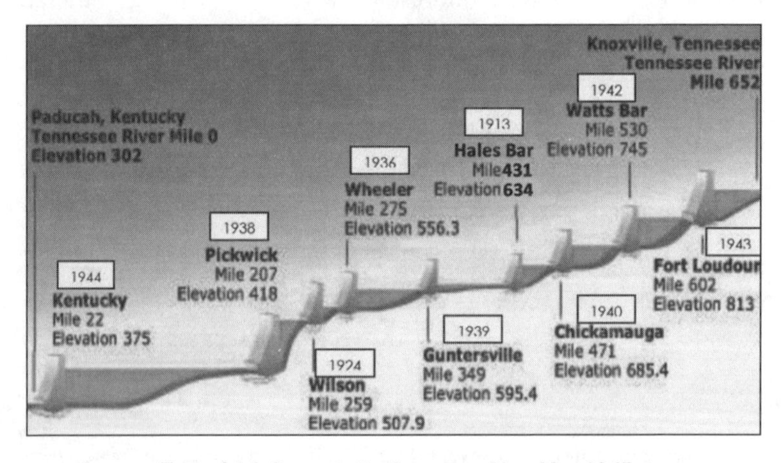

Figure 14.1: Dams on the Tennessee River After 1945

As we learned in Chapter Eight, Wilson Dam was constructed by the federal government and transferred to the TVA upon its formation. Hales Bar Dam was completed in 1913 below Chattanooga at the southwestern end of the Tennessee River Gorge by the predecessor of the Tennessee Electric Power Company. Before the dam and its lock had been constructed, the notoriously treacherous stretch of the river through the gorge had been a serious roadblock to navigation. TEPCO was Tennessee's largest investor-owned electrical power monopoly when the TVA was formed, and it filed suit to block competition from the TVA. In January 1939, the Supreme Court ruled in *Tennessee Electric Power Company et al v. Tennessee Valley Authority* that direct competition by the federal government in selling electricity was not unconstitutional. Soon thereafter, TEPCO sold Hales Bar Dam and three tributary dams on the Ocoee River in the mountains east of Chattanooga to the TVA.[3] (Hales Bar suffered from continuing leakage issues that led to its eventual replacement by Nickajack Dam in the 1960s.)

Soon after the TVA was established, Roosevelt directed it to start constructing Wheeler Dam, its first new main-stem dam, as quickly as possible in order to create jobs. The TVA was in the throes of getting organized, so it engaged the Bureau of Reclamation to prepare detailed designs and contract drawings for the dam. The Corps of Engineers designed and built Wheeler's lock. The dam began generating power in November 1936.

In the TVA's earliest days, Chairman Arthur E. Morgan, who also served as chief engineer, formed a consulting board of dis-

---

[3] The TEPCO sales package to the TVA also included Great Falls Dam, 75 miles southeast of Nashville on the Caney Fork River. It is the only TVA dam located outside the Tennessee River watershed. The dam was constructed in 1916 and had 31,860 kW generating capacity.

tinguished hydropower design and construction engineers to oversee the TVA's work. The consulting board made its first formal, general inspection of the progress of Wheeler's construction in May 1935 (see Figure 14.2). The extraordinary group of eleven men posing somewhat awkwardly was a veritable who's who of hydropower in the mid-1930s (see the Appendix for information about each member of the assembled group). It is amazing that the man on the far right, Lt. Col. Charles Perry of the Corps of Engineers, drowned less than three years later after falling from the top of the spillway wall at Wheeler.

Figure 14.2: TVA Consulting Board Inspection of Wheeler Dam
*Left to right: W.M. Hall; S.M. Woodward**; L. Evans; L.N. McClellan; C.H. Paul*; J.L. Savage*; C.H. Locher*; G.P. Jessup; C.A. Bock; L.F. Harza*; and Lt. Col. C.E. Perry.*
(consulting board members denoted by an asterisk)

The first tributary dam, Cove Creek Dam on the Clinch River, was specifically called for in the TVA Act. In honor of Senator Norris, it was renamed Norris Dam after passage of the Act, and its construction proceeded immediately. The next tributary dam to be built was Hiwassee Dam on the Hiwassee River in

North Carolina. Both were to provide critical flood protection for the Chattanooga area.

The approach of World War II added urgency to construction of additional tributary dams to meet vastly increased wartime electricity needs for essential aluminum production at Alcoa's Tennessee plants near Knoxville and for secret nuclear facilities being constructed at Oak Ridge. The TVA met the wartime challenge. In 1942, twelve dams, plus a steam plant, were being constructed simultaneously.[4]

TVA dams have a distinctive appearance. The TVA's first directors, especially Chairman Arthur E. Morgan, shared a utopian view of the agency's role—not only to harness the Tennessee River and produce cheap hydroelectric power but also to pursue a grander vision of broad regional planning and economic and social development. The dams were designed to enhance the TVA's image and to draw visitors and convey to them that the TVA was working for them to make their lives better. In an unusual move, the TVA gave visual appearance and public image primacy in the design process. The resulting dramatic, modernist designs were simple and functional, treating dam, powerhouse, and surroundings cohesively as a new kind of modern sculpture. Ornamentation was eschewed since it was reminiscent of structures built by the wealthy as monuments to themselves in the Gilded Age. The TVA built for the people in the Valley. The TVA remains the greatest modernist architecture project ever,

---

[4] Construction of two tributary dams in northeastern Tennessee—South Holston Dam on the South Fork Holston River and Watauga Dam on the Watauga River—was suspended late in 1942 when the projects were deemed not critical to the war effort. Construction resumed after the war. Watauga was completed at the end of 1948 (50,000 kW), South Holston in 1950 (35,000 kW).

melding an epic vision of social progress with sensitively executed structures on an unparalleled scale.

Two dams that were being built simultaneously—Fontana Dam and Kentucky Dam—illustrate the wide range of challenges the TVA faced in project design and construction. As World War II began, Alcoa's sprawling aluminum smelting facilities near Knoxville exhibited a seemingly insatiable need for electricity to supply aluminum for aircraft production. Alcoa originally had located there to take advantage of hydropower that could be developed from the Little Tennessee River as it cascaded downward from the mountains nearby. The TVA also had an interest in taming the Little Tennessee and had been eyeing Fontana as an ideal location for a significant hydroelectric facility.

Fontana is located 65 miles south of Knoxville, 61 miles up the Little Tennessee River. The river's headwaters are another 80 miles up into the Great Smoky Mountains. The immense annual rainfall within the watershed, coupled with deep, narrow gorges at Fontana, made it especially attractive for a dam whose reservoir could retain large amounts of water. The site was the only location within the Little River Valley where a dam of 400 feet or more in height was feasible.

The situation was complicated because Alcoa already had constructed two dams downstream from Fontana, and water releases from a TVA dam at Fontana would impact those downstream facilities. Alcoa's sudden urgent needs for electricity as war loomed led to a solution. In August 1941, the two parties signed an agreement whereby the TVA would have control over the water releases and electrical output of Alcoa's downstream dams and the TVA would guarantee that Alcoa would be the primary recipient of the dams' electricity. For the first time in US history, a privately owned generating system was to be regulated by a public agency per a negotiated agreement. Congress authorized construction of the Fontana Dam on an emergency basis ten days after Pearl Harbor was attacked, and construction began on January 1, 1942.

Fontana Dam is reminiscent of Bureau of Reclamation projects in the West. Like Hoover, it is sandwiched between the walls of a river gorge. At 480 feet in height, it is the tallest dam east of the Rockies. When completed, only Hoover and Grand Coulee were taller. Fontana would have the distinction of being the world's third tallest dam for a very brief time: The last concrete for Reclamation's 602-foot-tall Shasta Dam in California was poured on January 2, 1945, just about two months after construction on Fontana Dam closed.

Fontana is a concrete, straight gravity-type dam whose overall crest length is 2,365 feet. It is 375 feet thick at its base. Roughly 2.8 million cubic yards of concrete were used in the dam's construction, about two thirds as much as used for Hoover Dam. The six-story concrete powerhouse with dramatic glass walls on all sides is located at the base of the dam. There is not a typical spillway over the main face of the dam since it was feared that the water drop of more than 400 feet eventually would undermine the dam's foundation. Instead, spillways at the east end of the dam (right side facing the dam from downstream) feed two concrete-lined diversion tunnels, 34 feet in diameter. The diversion tunnels discharge into the river downstream past the powerhouse via a concrete bucket structure carefully designed to dissipate the energy of the emerging high-velocity water by deflecting it upward and toward the center of the river channel.

The east wing of the powerhouse serves as a visitor lobby and reception area. Due to the powerhouse's constricted space, TVA architects designed a separate visitor building at the top of the dam, with access to the powerhouse lobby via an inclined railway. From the attractive lobby with its marble walls and terrazzo floors, visitors enter a balcony overlooking the generator room floor and can view the operation of the generators below. They also can descend a staircase and view the control room through a plate-glass window. The interior south wall of the generator room includes lettering spelling out "1942—BUILT FOR THE PEOPLE OF THE UNITED STATES OF AMERICA—1945." The overall look of the dam is an excellent example of the modernist look that the architects wanted for TVA dams.

The construction of Fontana Dam presented many of the same challenges faced with tall western dams, as well as other challenges driven by the need for expedited completion to support the war effort. For starters, the dam site was in a remote mountainous area. Construction continued around-the-clock, seven days a week. The accelerated schedule required assembling a large work force in a short time and then housing that force. Both were huge tasks. Seven months after construction began, there were 5,241 workers on the project. Peak employment would reach 6,337 in July 1943. In order to house all these workers, it was necessary to quickly build a construction camp and a village with family housing, schools, a hospital, and other community features. Several temporary tent camps also were used. Feeding the workforce was a gargantuan undertaking. At the time of peak employment, the cafeteria daily served 7,000 meals and packed 2,000 lunches, with the cost deducted from workers' paychecks.

The number of laborers within commuting distance of Fontana was miniscule. Furthermore, stiff competition in the labor market and the rapid induction of men into the armed forces made it extremely difficult to attract and retain workers. Recruiting parties traveled far and wide to attract workers. House-to-house canvassing and newspaper and radio advertising were used extensively. One recruiting billboard proclaimed "Fontana Dam ... Power to Win." Another stated "Fontana Dam. Essential war project. Power for aluminum, aluminum for warplanes, warplanes for victory."

The TVA's Fontana project manager, Fred C. Schlemmer, worked hard to maintain the morale of the workers and their families in this remote location. Patriotic and big-band music blared from a system of loudspeakers around the work site. A sign at the entrance to the cafeteria exhorted "Work! Or Fight!" Schlemmer used the cafeteria's hearty food as a morale booster. He held mass employee meetings to sell war bonds, praise the work force for achieving construction goals, and promote teamwork. He also scheduled weekly dances, which were wildly popular.

Erecting Fontana Dam was a difficult task. As with Reclamation's tall western dams, dissipating the heat generated in the curing of the massive amounts of concrete poured to form the dam was critical. The main dam was built in 50-foot-wide blocks, with longitudinal contraction joints spaced 80–100 feet apart. Concrete was poured in 2.5-foot or 5-foot lifts. More than 500 miles of cooling pipes were embedded in the dam. Depending upon the season during which concrete was poured, river water or refrigerated cooling water was circulated through the cooling pipes. Despite all efforts, cracks were observed in the face of the dam starting in 1949. It was determined that the cracking was due to thermal expansion caused by a continuing chemical reaction between the cement and aggregate portions of the dam's concrete. A crack monitoring and control program remains in place today.

The TVA acquired 68,000 acres of land for the Fontana project, of which 7,726 were cleared and prepared for flooding. The terrain to be cleared was mountainous, rugged, heavily timbered, and nearly inaccessible—more so than in any other TVA project. The reservoir behind the dam was by far TVA's deepest and, in most areas, had steep shoreline slopes. The reservoir inundated six villages. As a result, 1,311 families were relocated, and 74 churches, stores, schools, mills, and mines were removed. The TVA knocked down and burned the houses, barns, and outbuildings of the people it displaced to keep them from returning. In addition, just over half of the 2,043 graves in forty cemeteries were moved to other cemeteries. There also were road, bridge, and railroad relocations. The reservoir formed behind the dam was approximately 29 miles long, with a shoreline length of around 245 miles.

The entire Fontana construction project was conducted at warp speed. The dam was completed on November 7, 1944, thirty-four months after it was started, and the reservoir began to fill. Fontana was completed in time to provide a much-appreciated power boost to Alcoa during the closing months of the war. The powerhouse was built to contain three 67,500 kW generating units. The first of those went into commercial operation on January 20, 1945, the second on March 24. With changing postwar priorities, the third generating unit was not placed in operation until 1954, then bringing Fontana's total power capacity to 202.5 MW.

More than 300 miles west at the opposite end of the TVA system, the Kentucky Dam was rising. It was the largest TVA project and entirely different from Fontana. Located in the rolling terrain of western Kentucky where the Tennessee, Ohio, and Mississippi rivers come together, Kentucky Dam is the spigot helping to control floods on the lower Ohio and Mississippi rivers and the gateway allowing for navigation along the entire length of the Tennessee River.

The Kentucky project had been contemplated even before the early days of the TVA. The Corps of Engineers 308 report[5] for Tennessee River development had identified a potential lower river dam project at Aurora Landing, 20 miles upstream from the Gilbertsville site where Kentucky Dam eventually was built. Both the Corps and the TVA at various times identified the project as the most important main-stem dam. The project's Congressional authorization, however, became caught up in a political drama.

This drama featured Arthur Morgan, TVA's Chairman—until he was fired by Roosevelt in 1938. It was Morgan who personally was in charge of the TVA river-development and dam-building program. Formerly the president of Antioch College in Yellow Springs, Ohio, he was an accomplished but idealistic and politically naive engineer.[6] He was dogmatic and

---

5 See Chapter Thirteen for more background on 308 reports.

6 Morgan (1878–1975) was a civil engineer and water control expert. Early in his career, he authored the Minnesota Water Control Code and served as the US Department of Agriculture's supervising drainage engineer. In 1910, he founded the Morgan Engineering Company, which specialized in flood plain drainage and reclamation. Three years later, after record Great Miami River flooding devastated Dayton, Ohio, Morgan was hired to develop a flood control plan for the river's entire 4,000-square-mile watershed. His innovative plan led to the formation of the Miami Conservancy District, with Morgan as its chief engineer. Before he became president of Antioch College in 1920, he oversaw the district's construction of an integrated system of five water storage basins fronted by massive earthen dams. The storage basins were dry and farmable except during flooding events. MCD was the first major watershed district in the United States, and the dams then were the country's largest embankment dams.

could accept no vision for the Tennessee River other than his own. Ever the visionary, starting early in 1934, Morgan secretly had his team begin investigating possibilities for a colossal dam to be placed on the Ohio River just above Paducah to simultaneously control the Ohio, Tennessee, and Cumberland rivers. In an internal memo, he said this "project would create one of the greatest stretches of inland navigation in the United States."[7]

He first considered a single 5.5-mile-long, earth-filled dam that would cut across both the Ohio and Tennessee rivers, thereby controlling those two rivers and the Cumberland River, which emptied into the Ohio a bit farther upstream. Simultaneously, he investigated a two-dam Ohio–Tennessee solution: one dam to be placed at Smithland, Kentucky, on the Ohio River and another at Gilbertsville, Kentucky, on the Tennessee River. Their reservoirs would be connected by a canal.

Morgan sought and received an opinion from the TVA's legal counsel that his Ohio River investigations were sanctioned by the TVA Act. In May 1935, he finally informed the other two TVA directors about his intensified, ongoing Paducah-area investigations and in June publicly acknowledged that the TVA was making Ohio River investigations—being vague about specifics in both cases.

Meanwhile, Morgan faced Congressional hearings on the TVA's budget authorizations for the coming fiscal year (fiscal 1936). He detested politicians and resented their oversight. Morgan already had alienated Kenneth McKellar, the powerful ranking member of the Senate Appropriations Committee, having sanctimoniously told him that politics would not be a factor

---

7 A.E. Morgan to C.A. Bock, Memorandum, February 5, 1935, TVA Corporate Library, p. 1.

within the TVA and dismissing him as another corrupt politician. McKellar reciprocated by saying that if the dam building program of the TVA succeeded, it would be in spite of Morgan "instead of by his help." Other key members of the House and Senate, including the chairman of the House Appropriations Committee, also already distrusted Morgan's handling of the lower Tennessee project.

The Congressional committees were expecting the TVA to be proceeding with a dam at Aurora Landing. Morgan faced a dilemma. He clearly favored the idea of building a single superdam across the Ohio at Paducah but needed time and money to conduct extensive field investigations to ensure the project could be constructed. In his Congressional testimony, he asserted that investigations in the Aurora Landing region were incomplete, that the final site in the Aurora area had not yet been determined, and that it was the TVA's "duty" to also investigate the superdam possibility.

Surprised to learn of Morgan's superdam plans, some Congressmen moved to rein him in and force commencement of work at Aurora. In the end, fiscal 1936 funds were authorized for "continuation of preliminary investigations as to the location and desirability of a dam at or near Aurora Landing."[8] The language was broad enough to allow continued investigations in the Paducah area on the Ohio River. An amendment to the original TVA Act also directed the Authority to report its recommendations for the entire river development program by no later than April 1, 1936. It is important to note that all of this transpired while other dramas raged, including pitched attacks on the TVA

[8] *Congressional Record*, 74th Congress, Session I, Ch. 508, August 12, 1935, p. 596.

by private electric utilities that were whipsawing Congress and a coup against Morgan by the TVA's other two directors.

With funding secured for fiscal 1936 and a deadline looming for submitting a river development plan, Morgan's engineers quickly began an intensive physical examination of potential dam sites on the lower Tennessee and on the Ohio in the Paducah area. By December 1935, they concluded that the Aurora Landing site on the Tennessee was unsuited due to unfavorable foundation conditions, that preliminary subsurface work downstream at Gilbertsville showed promise, that Dog Island on the Ohio would be a much better site than nearby Smithland for a two-dam solution, and that a large superdam plan at Paducah (aka the Paducah Plan) was technically feasible. Early in 1936, Morgan began promoting the Paducah Plan while also conceding that a two-dam solution (at Dog Island and Gilbertsville) with a connecting canal would retain a large portion of the Paducah Plan benefits, and that Gilbertsville was preferable to Aurora Landing for improvement for the Tennessee River alone (the TVA never admitted that the Aurora Landing project would have been impossible to build).

When Morgan submitted the formal presentation of the TVA's dam-building program, *The Unified Development of the Tennessee River System,* in March 1936, it outlined the three alternative recommendations for the Lower Tennessee: the Paducah Plan, the Dog Island/Gilbertsville two-dam solution, and the Gilbertsville location for a single-dam, Tennessee-River-only solution. It was obvious to key Representatives and Senators that Morgan hoped to build the Paducah Plan project, and they swiftly lined up in opposition. They realized that the Paducah Plan would be way too expensive and that building on the Ohio River was outside the TVA's legal mandate. Congress authorized

$650,000 for fiscal 1937 for continuing investigations on the Tennessee alone, expecting that planning phases for Gilbertsville would proceed at an accelerated pace.

Nature intervened in the dam drama. In January 1937, heavy, persistent rains along the length of the Ohio River Valley led to the third deadliest flood in US history. As a result of the flood, 385 people died, and roughly a million were left homeless. Property losses reached $500 million (more than $9 billion in today's dollars). At Paducah, the home of Alben Barkley, soon to become Senate majority leader and later President Harry Truman's vice president, 18 inches of rain fell in 16 days. Nearly 95 percent of the city was submerged, and 27,000 of the city's 33,000 total residents were forced to flee to higher ground. The flood was the worst natural disaster in Paducah's history.

The disaster focused intense attention on flood control in the lower Tennessee Valley and revived the possibility of a Paducah superdam. As Congressional hearings began in April 1937 on the TVA's appropriations for fiscal 1938, Morgan continued to favor his superdam, saying the dam "would probably be the largest construction project in America; it would be ... of greater importance than the Boulder Dam. It would be almost the major dam construction project in flood control in America." He reluctantly estimated that it would cost "about $300 million."[9]

The projected cost shocked members of Congress: Cumulative appropriations for all TVA construction activities through fiscal 1938 totaled $191 million. Morgan's superdam would cost at least 1.6 times that total amount. Despite his Paducah Plan

---

9 B. Anthony Gannon, "Vision or Obsession? Arthur E. Morgan and the Superdam," *The Register of the Kentucky Historical Society,* Vol. 97, No. 1, Winter 1999, p. 78.

aspirations, Morgan did go on to state that he was satisfied that the Gilbertsville Dam was a proper element for TVA construction since it also later could be part of his potential two-dam solution, and he requested more money for its further planning. The total estimated cost for the Gilbertsville project was around $110 million.

Senators Barkley and McKellar had had enough. They pushed through legislation that, in May 1937, authorized construction of the Gilbertsville (or Kentucky) project. Over the next year, detailed project plans were undertaken, and land acquisition began. Preliminary construction began July 1, 1938. Meanwhile, Morgan found himself no longer able to withstand attacks from Barkley and McKellar and from the TVA's other two directors—nor could he escape Roosevelt's ire. FDR fired Morgan for "contumacious" (i.e., stubbornly or willfully disobedient) behavior three months before construction on the Kentucky Dam began at Gilbertsville.

The design for the Kentucky Dam project was driven by several factors. The height of the dam was set to create a reservoir stretching 184 miles upstream to Pickwick Dam with at least a 9-foot channel. Whereas Fontana was the tallest TVA dam, Kentucky Dam would be the longest since it had to stretch across the entire 1.6-mile-wide flood plain at Gilbertsville. To drive all cofferdam piling to the area's deeply buried bedrock, the maximum cofferdam height would be an unprecedented 100-plus feet. Because the dam site was only 75 miles from the epicenter of the 1811–12 New Madrid earthquakes (the most powerful ever to hit the United States east of the Rockies), it was imperative to design the dam to withstand earthquake shock.

There also needed to be a navigation lock that could handle large river barges. Furthermore, the Illinois Central Railroad's main Chicago-to-New-Orleans line crossing the river at the dam site would have to be relocated.

Original plans for the Kentucky project did not include hydroelectric power generation, although the designs did allow for a future powerhouse. Escalating power needs soon led to plan revisions adding power generating capability to the initial construction.

Kentucky Dam's total length is 8,422 feet. The 24-gate, concrete, northward-facing spillway straddles the river channel and adjoins the powerhouse, which holds five 32 MW generators. Earth-filled embankments stretch over the flood plains to the west and to the east. The navigation lock is near the end of the east embankment. The switchyard structure is located downstream between the lock and powerhouse. The single-lift navigation lock, the only lock on the Tennessee River designed by the TVA (all others by the Corps), has clear chamber dimensions of 110 feet wide by 600 feet long and a maximum lift height of 73 feet. The relocated Illinois Central rail line and a state highway traverse the top of the 206-foot-tall dam. Due to the depth of the foundation rock, the dam is more than half submerged and appears much less massive than it actually is.

The dam's construction required 1,356,000 cubic yards of concrete, 6,140,000 cubic yards of excavation, 5,064,000 cubic yards of earth fill, and 518,000 cubic yards of riprap. Building Kentucky Dam took six years from the start of construction until the reservoir began filling at the end of August 1944. At the peak of construction, nearly 5,000 men were at work building the dam and preparing the reservoir area.

In many ways, Kentucky Lake, the 184-mile-long reservoir behind the dam, is what ultimately defined the Kentucky project. It is the largest artificial lake by area in the eastern United

States. It has a volume of more than 6 million acre-feet, about 260,000 acres of water surface when filled to the top of the dam's gates, and 2,380 miles of cove-studded shoreline. The dam and lake help reduce flood damage for 6 million acres of the lower Ohio and Mississippi rivers and curb the frequency of flooding on another 4 million acres.

Preparing the reservoir required surveying and mapping about 685,000 acres of land, purchasing 320,000 acres, and clearing 49,000 acres. Several communities were inundated in order to make the project reality. A protective dike was constructed at one town to protect it from reservoir backwaters. Roughly 2,600 families, 3,390 graves in 397 cemeteries, and 365 miles of highways and railroads had to be relocated. The project featured 65 new bridges, 7 rebuilt bridges, and 3 raised bridges.

George Jessup, the manager on Conowingo Dam and the TVA's Wheeler and Guntersville dams, also served as project manager for the Kentucky Dam project. Jessup made frequent reservoir site inspection trips by boat. He also pioneered the use of aerial photography as reservoir mapping and preparation proceeded. Jessup was known for constantly clenching a curved-stem, Sherlock Holmes-style pipe in his teeth and wearing a signature broad-brimmed hat. He was careful not to alter his appearance whenever he ventured upriver so that moonshiners emerging along the banks would not mistake him for a "revenuer" and shoot him on sight.

Upon completion of Kentucky Dam construction in August 1944, the final link was formed in the navigation channel stretching the entire length of the Tennessee River. Lock operations began immediately, but it took more than eight months to fill the reservoir enough to create a 9-foot channel all the way to Pickwick Landing Dam. Then, for the first time, deep-draft commercial freight traffic could move freely year-round over the entire river.

The lock at Kentucky Dam literally became the Tennessee Valley's doorway to the 21-state Inland Waterway System and the world beyond. Its completion was a major factor in stimulating the Valley's economic development. Freight traffic on the Tennessee River has grown from 2 million tons of cargo a year upon the completion of the lock and dam in the 1940s to around 50 million tons today. The lock handles more than two thousand loaded barges a month. A new, adjoining, larger lock (twice as long as the existing lock) now under construction will be better able to accommodate today's long barge tows and will significantly lessen lock transit delays.

As important today as it was when construction was completed in 1944, Kentucky Dam remains an engineering marvel. Five weeks after World War II ended in 1945, President Harry Truman dedicated Kentucky Dam before the largest crowd ever assembled in western Kentucky. Among the attendees were dignitaries such as Senator Barkley, fifteen members of Congress, George Marshall, Dean Acheson, Averell Harriman, and TVA Chairman David Lilienthal. In his speech, Truman extolled the TVA and advocated creation of similar regional agencies elsewhere:

> The completion of this dam marks a new high point in modern pioneering in America ... But (the TVA) is more than dams and locks and chemical plants and power lines. It is an important experiment in democracy ... Why has TVA succeeded so well? Why does it have the esteem of the people of this Valley, and attract the attention of other regions of America, and of the entire world? To me the answer is clear—TVA is just plain commonsense. It

> is commonsense hitched up to modern science and good management. And that's about all there is to it ... It is easy to see that most of these commonsense principles can be applied to other valleys, and I have already recommended to the Congress that a start be made in that direction. Careful planning and commonsense development can convert the idle and wasting resources of other valleys into jobs and better living ...[10]

After the ceremony, Truman and Jessup had a private conversation. Jessup, like Truman, was direct, gruff, and plain-spoken. During the conversation, "they cussed each other out."[11] It seems Jessup called Truman a "damned fool" because Truman had made absolutely no reference to the thousands of workers who had built the TVA dams (many of whom had been standing there listening to Truman's speech) and uttered no thanks or appreciation to them for what they had accomplished. Truman's reply apparently was equally fiery.

It has been said that the Fontana project manager, Fred Schlemmer, and the Kentucky project manager, George Jessup, were the men who built the TVA.[12] They also have been referred to as "TVA East" and "TVA West." The two men collectively managed the construction of seven of the sixteen hydroelectric dams built by the TVA by the end of World War II. Schlemmer's dams were Norris, Watts Bar, Chickamauga, and Fontana. Jessup's were Wheeler, Guntersville, and Kentucky.[13]

---

[10] National Archives at Atlanta, Record Group 142, National Archives Identifier 281490.

[11] Private conversation between Jessup and the author, 1960.

[12] See, for example, *The Knoxville Journal,* May 23, 1943, p. 31.

[13] There's an it's-a-small-world story about these two men: My grandfather was George Jessup. When my brother met the woman who would become

Although the TVA generally received praise for its accomplishments through the Great Depression and World War II, it did have its detractors in the Tennessee Valley. The TVA was viewed by many as an occupying force—few key employees were from the Valley. The forced relocation of landowners by TVA projects led to strong resentment. Many families had been living on the land for generations. The TVA purchased property from willing sellers as often as possible, but it also proceeded with condemnation when owners resisted. The prevailing feeling was that the TVA underpaid owners and roughly evicted them. The many sharecroppers in the region typically got nothing and suddenly had nowhere to work. The best farmland in the vicinity of TVA dams and reservoirs often ended up under water. In little more than a decade, the TVA displaced 70,000 citizens from 467,000 acres of prime farmland.[14] The reservoir issue would prove to be a major impediment to the expansion of hydropower in years ahead.

The remarkable total of sixteen hydroelectric dams constructed by the TVA between its formation in 1933 and the end of World War II is detailed in Figure 14.3. Seven of the dams were constructed on the Tennessee River itself and nine on tributaries. By 1946, the TVA provided electricity to 668,752 consumers and had become the nation's largest integrated electricity supplier.[15] Three quarters of the Tennessee Valley had electrical service. By the early 1950s, electrification was nearly universal throughout the Valley.

---

his wife, he was surprised to learn that she was the granddaughter of Fred Schlemmer.

[14] Steven M. Neuse, *David E. Lilienthal: The Journey of an American Liberal,* Knoxville, TN: The University of Tennessee Press, 1996, p. 139.

[15] See, e.g., https://tennesseeencyclopedia.net/entries/tennessee-valley-authority/.

Dam	Location	Year Completed	Power Capacity (kW)	Comments
Tennessee River				
Wilson	Mile Marker 259, AL	From USACE, 1933	436,000	See Chapter Seven.
Wheeler	Mile Marker 275, AL	1936	259,200	Built to extend 9-ft slack-water channel 74 miles through shoal area above Wilson Dam to Guntersville.
Pickwick Landing	Mile Marker 207, TN	1938	216,000	63-ft single-lift lock height (greater than Bonneville).
Guntersville	Mile Marker 349, AL	1939	97,200	Provides 9-ft channel upstream through a section of the river that had presented serious obstacles to navigation.
Hales Bar	Mile Marker 431, TN	From TEPCO, 1939	99,700	Replaced by Nickajack in 1967 due to continuing leakage.
Chickamauga	Mile Marker 471, TN	1940	108,000	Located just above Chattanooga to provide flood control.Difficult foundation conditions. Dam is 129 ft tall, 5,794 ft long.
Watts Bar	Mile Marker 530, TN	1942	150,000	Numerous shoals and bars above dam's location. 73-ft lock lift.TVA's first steam plant also built here, operational in 1942.
Fort Loudoun	Mile Marker 602, TN	1943	128,000	80-ft single lift lock.Project expanded and accelerated due to wartime power needs. Water from Little Tennessee River diverted through short canal to reservoir for power production.
Kentucky	Mile Marker 22, KY	1944	160,000	Gateway to Ohio/Mississippi Rivers.Key for flood control.
Tributaries				
Norris	Clinch River, TN	1936	100,800	Originally called Cove Creek.Identified in TVA Act.To regulate flow into entire main stem of Tennessee River.
TEPCO Purchases				
• Ocoee #1	Ocoee River, TN	From TEPCO, 1939	18,000	Construction completed in 1912.For Chattanooga power.
• Ocoee #2	Ocoee River, TN	From TEPCO, 1939	19,900	Completed in 1913.5-mile flume to powerhouse.
• Blue Ridge	Toccoa/Ocoee River,GA	From TEPCO, 1939	20,000	Completed in 1931.23 miles upstream from Ocoee #3.
Hiwassee	Hiwassee River, NC	1940	57,600	Dam 307 feet tall.Construction techniques similar to Reclamation dams to prevent cracking.Nation's first reversible pump/turbine installed here in 1956, adding 59,500 kW power capacity.
Chatuge	Hiwassee River, NC	1942	----	Earth and rock dam originally for headwater storage, with provision for 10,000 kW power unit added in 1954.
Cherokee	Holston River, TN	1942	135,200	First major National Defense Emergency Program project for power to support war effort.Completed in 16 months.

Dam	Location	Year Completed	Power Capacity (kW)	Comments
Tributaries (continued)				
Nottely	Nottely River, GA	1942	----	Part of Hiwassee River System project. Earth and rock dam for flow regulation.15,000 kW generator added in 1956.
Apalachia	Hiwassee River, NC	1943	75,000	Built on emergency basis to support WW II aluminum production.Dam is 150 ft tall.
Douglas	French Broad River, TN	1943	112,000	Completed in less than 13 months to supply electricity for war effort.
Ocoee No. 3	Ocoee River, TN	1943	27,000	Construction expedited to produce power for war effort.Powerhouse over 2 miles downstream from dam.Land purchased from TEPCO in 1939.
Fontana	Little Tennessee River, NC	1944	202,500	Highest dam east of the Rockies.Constructed in record time in wilderness area to provide power for wartime aluminum production.Over 5,000 workers housed.

Table excludes Great Falls Dam, which lies outside the Tennessee Valley.

Figure 14.3: Tennessee Valley Authority Hydroelectric Dams in 1945

The TVA was a key component of President Franklin Roosevelt's New Deal. Roosevelt used his New Deal to move the federal government to center stage as a starring player in the economic life of the country. Soon after becoming president in 1933, he formed the TVA to undertake centralized planning for and development of the Tennessee Valley.Multipurpose projects were to be the means by which to control and develop the water-related resources of the entire Tennessee River system and to put people to work during the Depression, thereby advancing the Valley's social and economic well-being. Despite this lofty goal, by the end of World War II, the TVA had morphed into being little more than a giant, waterpower-driven, public utility with a Valley-wide monopoly.

In 1945, the TVA was approaching the limits of its hydropower production possibilities. Its boom years of dam building were over. The Tennessee River was dammed all the way from Knoxville to Paducah. The hydropower potential of the river's eastern tributaries was largely tapped. Two tributary dams begun in 1942 but suspended due to wartime priorities were completed by 1950. Two more hydroelectric dams on the South Fork River near Kingsport, Tennessee, were the only other dams built before the 1960s.

Postwar power demands continued to grow, especially due to the needs of federal institutions such as the Atomic Energy Commission. By the late 1950s, more than half of all TVA power was consumed by federal agencies. The TVA began constructing coal-fired steam plants to satisfy the demand. After 1955, steam-plant electricity generation exceeded waterpower production. By the mid-1960s, the TVA was the nation's largest consumer of coal, much of it from strip mines in eastern Kentucky. Twenty years later, less than 10 percent of the TVA's total power generation was hydropower.

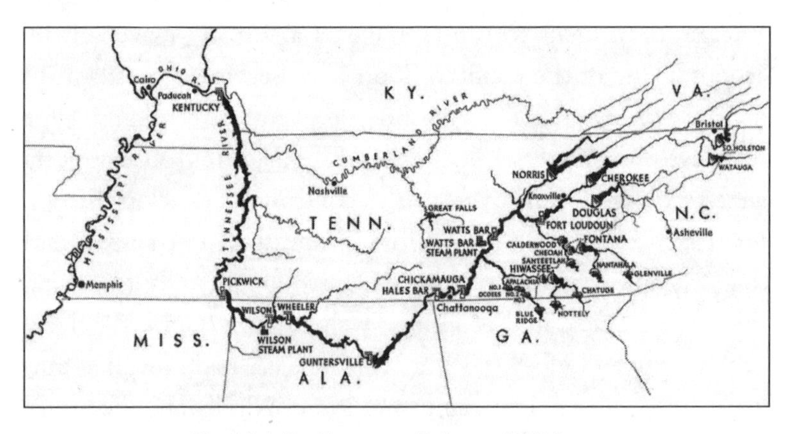

Fig. 14.4. The Tennessee Valley and TVA Dams

Fig. 14.5. Farm Family Near the Norris Dam Site in 1933

Fig. 14.6. F. Schlemmer, Senator G. Norris, and A. E. Morgan (L to R) at Norris Dam Construction Site

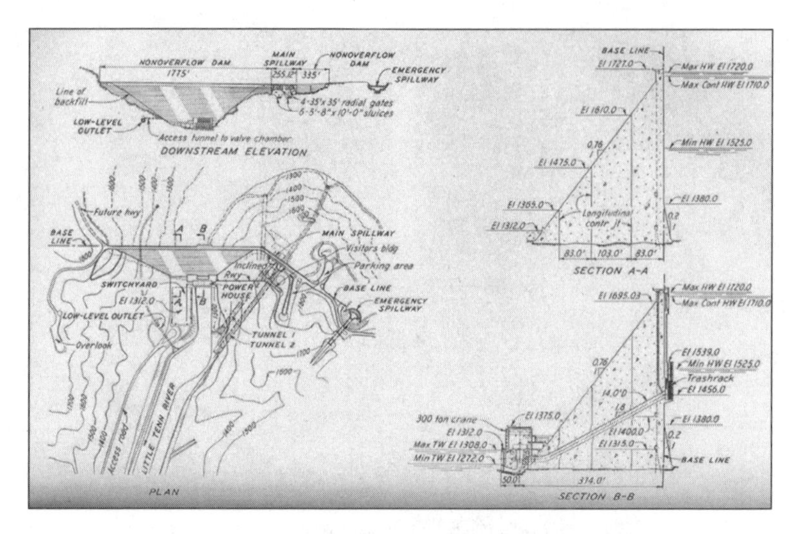

Fig. 14.7. Fontana Dam General Layout

Fig. 14.8. Fontana Dam Nighttime Construction

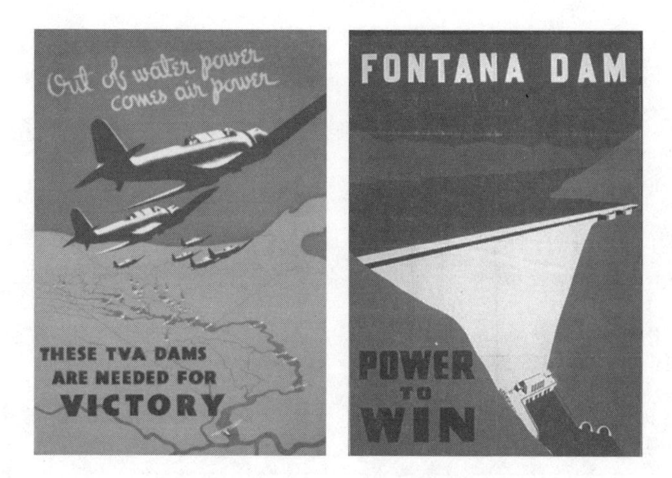

Fig. 14.9. TVA Recruitment Posters

Fig. 14.10. Fontana Dam upon Completion

Fig. 14.11. 1937 Paducah Flood

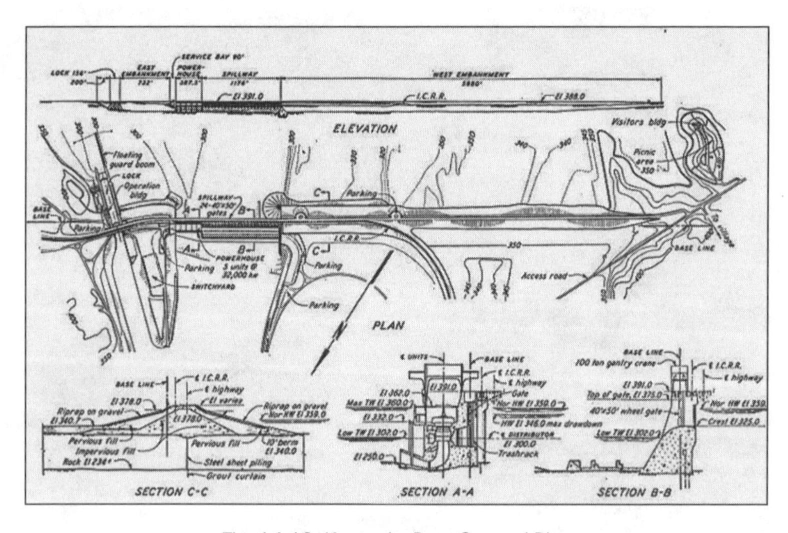

Fig. 14.12. Kentucky Dam General Plan

Fig. 14.13. Kentucky Dam in 1946

Fig. 14.14. Powerhouse and Gantry Crane, Kentucky Dam

Fig. 14.15. Lock at Kentucky Dam

Fig. 14.16. G. Jessup, Truman, and TVA Directors J. Pope and D. Lilienthal (L to R) at Dedication of Kentucky Dam

Fig. 14.17. Truman Dedicating Kentucky Dam (G. Jessup Second from Left)

Chapter Fifteen

Vanquished

Breathtaking progress was made in electrifying America in the half century after Thomas Edison began lighting New York City's financial district in 1882. By 1932, 70 percent of homes in the United States had electricity. From the earliest days, entrepreneurial enterprises dominated the industry. Many of the towering figures leading these companies still were at the helm as the 1930s began. They symbolized a deeply rooted American private tradition.

The growth of electricity until the 1930s largely was hydropower-driven. Where it was available, waterpower was a cheaper energy source than coal. Hydropower consistently represented 35–40 percent of the country's total annual electricity generation from 1910–32. More hydropower plants were built in the 1920s than in any other decade.

Electric companies controlled all functions in bringing power to their customers: generation, transmission, distribution, and marketing. They provided what became an essential service to society and had a natural monopoly. State government regulation became the norm.

In the early 1900s, electric utilities began merging to obtain territorial dominance, economies of scale, and better financing. This naturally led to the formation of holding companies by Insull, Stone & Webster, and others to facilitate management and financing of multiple separate utilities. By the 1920s, with the industry rapidly consolidating physically and financially, holding companies were formed to own other holding companies. Largely unregulated interstate holding companies emerged. Before long, some multilevel holding companies had more than five levels.

As demand for electricity skyrocketed, private utility companies gobbled up ideal dam sites for new hydropower facilities and soon were controlling rivers in addition to electricity markets.

The electricity-using public found itself serviced by a small number of interstate holding companies. In 1932, the eight largest holding companies controlled about 75 percent of the investor-owned electric business.[1] A handful of people in private industry had gained control over important societal needs of the entire nation.

A Federal Trade Commission investigation begun in 1928 uncovered holding company abuses, including overvalued assets, excessive fees charged for services provided to operating companies, insider trading, and other financial wrongdoing. Public antagonism toward holding companies intensified after the stock

---

[1] Leonard S., Andrew S., and Robert C. Hyman, *America's Electric Utilities: Past, Present, and Future, Eighth Edition,* Vienna, VA: Public Utilities Reports, Inc., 2005, p. 139.

market crash of 1929 decimated the value of holding company securities. The collapse of Insull's holding companies wiped out the life savings of six hundred thousand shareholders.

The battle between private power and public power became brutal in the 1930s during Franklin D. Roosevelt's presidency. Roosevelt was a strong supporter of public power. While New York governor, he established the New York Power Authority to develop hydropower on the St. Lawrence River. During his 1932 presidential campaign, he championed the reformation of the electric industry. He maintained that electricity had become a necessity deserved by all Americans, it had grown into an interstate business of vast proportions, and it required strict federal regulation and control. He asserted that the electric utility industry was dominated by a few self-interested behemoths who were fleecing the public. He decried the fact that only 10 percent of farms had electricity.

Soon after becoming president in 1933, Roosevelt authorized the construction of Grand Coulee and Bonneville dams on the Columbia River. He deftly justified these federal government projects in terms of construction job creation and regional economic development. He was silent about how to market the hydropower that the dams would generate. Although the private power industry objected to these projects, it was the nearly simultaneous creation of the TVA that unleashed the industry's wrath. It saw the TVA as a direct threat.

Samuel Insull had been the private utility industry's leader and spokesperson until his downfall in 1932. A new leader emerged to spearhead the fight against the TVA: Wendell Willkie. Willkie, a lawyer born in a small Indiana town, had become an attorney with Commonwealth and Southern in 1929. C&S, the country's third largest utility holding company, controlled the Alabama Power Company and the Tennessee Electric Power Company. Three months before Roosevelt became president, forty-year-old Willkie was named president of C&S.

Willkie commanded attention in the courtroom, the board room, and in public gatherings. He was more than six feet tall and solidly built, had a shock of tousled dark hair, and wore rumpled suits. He was eloquent and charming. He also was cocksure and impetuous.

It was this man who appeared as a witness at a Congressional hearing after the legislation to form the TVA was introduced. Willkie stated that he was not opposed to government power production if the government built no transmission lines to market the power. Roosevelt, however, insisted that the legislation include language authorizing them. The TVA Act additionally called for the TVA to give preference to municipal utilities and cooperatives in selling its power.

Upon passage of the Act, Roosevelt appointed the TVA's three initial directors. Arthur E. Morgan, a respected but idealistic civil engineer and college president we met in Chapter Fourteen, was named director and chairman of the Authority. The other directors were Harcourt A. Morgan (no relation to Arthur), the president and former dean of agriculture at the University of Tennessee, and David E. Lilienthal. Lilienthal, a thirty-three-year-old Harvard Law School graduate and utility law specialist, headed the Wisconsin Public Service Commission.

Arthur Morgan and Lilienthal clashed from the instant they arrived at the TVA. At the first meeting of the three directors, Arthur Morgan reported that he had heard from Willkie and had assured him of TVA cooperation. Morgan proposed that they negotiate the transfer of a small number of urban and rural C&S areas to the TVA and that a panel of economists devise methods to compare TVA operations of those yardstick areas with areas operated by private utilities. Lilienthal was aghast. He distrusted private utilities and believed that public power should be distributed publicly, through a network of local municipal power boards and rural cooperatives.

Harcourt Morgan and Lilienthal also concluded after that first meeting that Arthur Morgan was a terrible administrator. The two formed an alliance. At the next meeting of the three directors, they forced a division of responsibilities. Arthur Morgan would be responsible for engineering projects and several social experiments, Harcourt Morgan for agriculture and forestry, and Lilienthal for power and legal matters.[2]

Soon thereafter, Lilienthal and Arthur Morgan were summoned to meet with President Roosevelt in an attempt to resolve their differences about how widely to distribute TVA power. Roosevelt told them to work it out. Lilienthal promptly issued a TVA press release broadsiding the region's existing private utilities. It declared:

> The business of generating and distributing electric power is a public business. ... The interest of the public in the widest possible use of power is superior to any private interest. Where the private interest and this public interest conflict, the public interest must prevail. ... The right of a community to own and operate its own electric plant is undeniable. This is one of the measures which the people may properly take to protect themselves against unreason-

---

[2] The relationship between Lilienthal and Arthur Morgan continued to deteriorate. In 1936, Morgan told Roosevelt that he adamantly opposed Lilienthal's reappointment as a TVA director. Lilienthal began actively undermining Morgan, ultimately resulting in Roosevelt's firing of Morgan in 1938.

> able rates. Such a course of action may take the form of acquiring the existing plant, or setting up a competing plant. ... The authority cannot decline to take action solely upon the ground that to do so would injure a privately owned utility.[3]

It went on to say that the TVA's territory was contemplated to ultimately include the entire drainage area of the Tennessee River in Tennessee, Kentucky, Alabama, Georgia, North Carolina, and northeastern Mississippi. Based upon special considerations, it might extend even further. Willkie raged that to "take our markets is to take our property." TVA opponents screamed that the federal government was pushing into socialism.

Then, in September 1933, Lilienthal announced the TVA's proposed rates. The TVA would charge 3 cents per kilowatt hour for the first 50 kilowatts, gradually decreasing to 0.4 cents for usage over 400 kilowatt hours. Private rates in the area ranged from 4.6–5.8 cents.[4]

There was a reason for the timing of Lilienthal's announcements. Ownership of Wilson Dam was transferred from the Corps of Engineers to the TVA upon its formation. This transfer included the federal government's contract to sell Wilson's hydroelectricity exclusively to Alabama Power until January 1934.

Lilienthal and Willkie met for the first time at the beginning of October 1933 to negotiate terms for contract renewal. Lilienthal recalled that "we were two exceedingly cagey fellows."[5] The

---

[3] "Tennessee Power Held Public Right," *The New York Times,* August 25, 1933, p. 23.

[4] Neuse, *David E. Lilienthal,* p. 80.

[5] Thomas K. McCraw, *TVA and the Power Fight, 1933–1939,* Philadelphia, PA: J.B. Lippincott Company, 1971, p. 63.

two did need to reach an agreement. The TVA could not use all of the hydropower generated by Wilson Dam, then its only source of power revenue. It also needed to begin servicing customers other than Alabama Power in order to have a yardstick for comparison with private utility operations. Willkie wanted to contain the TVA to as small a service area as possible and avoid direct competition within service territories.

For the next two months, the two men cautiously circled each other looking for common ground. Pressure on Willkie increased as the TVA authorized building a transmission line between Wilson and Norris dams and signed an agreement to sell power to the municipal utility in Tupelo, Mississippi. Willkie and Lilienthal eventually reached an agreement, which was signed on January 4, 1934. They agreed that Willkie's companies would sell certain properties in northwest Alabama, northeast Mississippi, and eastern Tennessee to the TVA; that the TVA would continue selling Alabama Power the surplus power from Wilson Dam; that C&S would not sell electricity within the ceded area; and that the TVA would forego further incursions into the private companies' markets. The contract would last for five years or until six months after Norris Dam began producing power, whichever came first. Willkie told reporters, "It's tough to negotiate with a pistol at your head."

The battle between the TVA and C&S was just beginning. Willkie and Lilienthal continued to spar for the next five years, employing delays, blatant propaganda, litigation, and territory poaching. The two men distrusted each other. Lilienthal believed Willkie and his allies were devious obstructionists. Willkie thought Roosevelt and Lilienthal intended to destroy private utilities and make their shareholders' investments worthless. Lilienthal used substantial subsidies from Roosevelt's Public

Works Administration and Rural Electrification Act funds to aggressively promote the formation of municipal utilities and rural cooperatives to take business from the private utilities in the region.

In September 1934, a small group of Alabama Power shareholders brought suit against Alabama Power and the TVA, charging in *Ashwander v. Tennessee Valley Authority* that the TVA was unconstitutional and seeking to void the January 4 contract signed by Willkie and Lilienthal. A federal district court judge issued an injunction annulling the contract and prohibiting Alabama communities from accepting PWA funds to build distribution systems while his decision was being appealed. In July 1935, the federal circuit court overturned the injunction, and the case went to the Supreme Court. In February 1936, the Supreme Court ruled that the TVA had the right to acquire transmission lines and sell whatever power it produced at Wilson Dam. It did not rule on the TVA's constitutional status.

Three months later, nineteen utilities, led by C&S's Tennessee Electric Power Company subsidiary, filed suit claiming that the TVA was unconstitutional and seeking an injunction against its power program. In December 1936, a district court judge issued an injunction that blocked for six months the TVA from extending its service area or power lines or from starting new construction. In January 1938, the district court upheld the constitutionality of the TVA Act and decided the factual issues of the TEPCO case in the TVA's favor. The case was appealed to the Supreme Court. It ruled a year later that the utility companies lacked standing to file suit because they never had been guaranteed a monopoly.

Willkie saw the handwriting on the wall by early 1938 and began intense negotiations with Lilienthal for the sale to the TVA of all TEPCO properties and of small parts of Alabama and Mississippi Power. The two men eventually agreed on a $78.6 million settlement (roughly $1.5 billion in today's dollars). After the transaction received Congressional approval, the sale was formally completed on August 15, 1939. Public power ruled the Tennessee Valley.[6]

While the TVA battle raged, Roosevelt simultaneously pursued his crusade to deal with utility holding companies. He established the Securities and Exchange Commission in June 1934 to regulate securities transactions. In November, he met with Lilienthal and several other advisers to plot a strategy to destroy utility holding companies. In his State of the Union address in January 1935, he called for the "abolition of the evil of holding companies." In a meeting three weeks later, Willkie asked, "Do I understand then that any further efforts to avoid the breaking up of utility holding companies are futile?" Roosevelt gave him one look and shot back, "It is futile."[7]

On February 6, legislation for the Public Utility Company Act of 1935 was introduced in Congress. The legislation required all interstate holding companies to register with the SEC. After January 1, 1938, the SEC would be able to eliminate any holding companies that still controlled more than one geographically integrated system (generally thought to mean interstate

---

6 Lilienthal became chairman of the TVA in 1941, and the first chairman of the Atomic Energy Commission in 1946. Willkie was the 1940 Republican nominee for president. After the election, Willkie provided vital support to Roosevelt's wartime diplomacy. Roosevelt and Willkie talked shortly before Willkie's unexpected death in 1944 about joining forces in the future to form a new, progressive political party.

7 David E. Lilienthal, *The Journals of David E. Lilienthal, Volume 1, The TVA Years*, New York, NY: Harper & Row, 1964, p. 47.

systems). The elimination proviso quickly was dubbed the "death sentence." The legislation also gave the Federal Power Commission regulatory authority over interstate electricity sales.

Industry opposition to the legislation was blistering. Masses of small utility stockholders wrote letters to their congressmen. The House was flooded with eight hundred thousand letters and telegrams of opposition. The scale of industry lobbying was unprecedented. Roosevelt's House floor leader told him that the lobbying was the richest and most ruthless Congress had ever known. Willkie put on masterful performances as a witness at a House hearing on the bill and in articles, speeches, and country club talks. He focused his opposition primarily on the brutality of the so-called holding company death sentence.

The bill passed the Senate by a single vote. It became clear that there were not enough votes in the House for the death sentence. Despite heavy-handed pressure from Roosevelt, the House passed a version of the bill with the death sentence stripped out. This was Roosevelt's first major House defeat. After massive maneuvering, a compromise was reached by the conference committee established to reconcile the House and Senate versions of the bill. Holding companies could control more than one system if the additional system was too weak to stand alone and it was not too large or dispersed to prevent efficient management or regulation. No more than two holding companies would be allowed above an operating utility. Roosevelt signed the Act in August 1935.

Roosevelt's final move in reshaping the private power industry was in the Pacific Northwest. He envisioned creating TVA-like authorities in other parts of the country. As Bonneville Dam construction was nearing completion, with the TVA battle working its way through the courts, he was unable to build enough support in Congress to create a Columbia Valley Authority. Instead, the Bonneville Project Administration was created in August 1937 to market Bonneville's hydropower and build transmission lines to deliver it to utilities who would, in turn, distribute it to retail customers. Preference was given to public bodies and cooperatives in delivering electricity. The BPA became part of the Department of the Interior, and the Corps continued to operate the dam.[8]

Roosevelt expected the BPA to soon be replaced by a comprehensive TVA-like authority. In fact, the BPA's enabling legislation stated that the agency was "intended to be provisional pending the establishment of a permanent administration for Bonneville and other projects in the Columbia River Basin." In 1940, the BPA's name was changed to the Bonneville Power Administration, and the sale and transmission of power from Grand Coulee Dam were added to its charter. As other federal dams in the Columbia River basin were built, they also were added to the BPA's jurisdiction. Despite recurring efforts by Roosevelt and, subsequently, by President Harry Truman, political opposition and the private power lobby blocked formation of a CVA. Today, the federal government continues to operate the BPA as well as three other regional power marketing administrations.

Roosevelt's relentless New Deal push for public power radically changed the electric utility landscape. His power initiatives were aimed at simultaneously breaking up utility trusts, reasserting public use and development of remaining prime dam sites, and electrifying rural areas. His New Deal inducements

[8] The first administrator of the BPA was Seattle City Light's J.D. Ross. He repeatedly threatened Stone & Webster and Puget Sound Power, pressuring Puget Sound Power to sell its properties to public utility districts in the region.

to electrify America's farms succeeded. By 1949, 78 percent had electricity.[9] Within five more years, rural electrification was almost universal. America was supercharged.

The Public Utility Holding Company Act turned the private utility industry upside down. Operating control reverted from holding company financial tycoons to locally operated utilities focused on customer service. C&S ceased to exist, but Alabama Power continued to thrive. Insull's empire dissolved, but Chicago's Commonwealth Edison prospered. Between 1935 and 1950, 759 companies were separated from holding company systems.[10]

Through the TVA and the Bonneville and Grand Coulee dam projects, Roosevelt focused on developing basin-wide plans to tame the country's major waterways. Flood control, irrigation, abundant electricity, and economic development were important societal goals. This vision became widely accepted and gave federal dam construction a major advantage over private-utility dam construction. The federal government could allocate a significant portion of a dam's construction cost to its non-hydropower benefits, with associated costs to be borne by taxpayers. In order to obtain project financing, private utilities had to justify their projects solely on the payback from the hydropower to be generated.

Roosevelt's initiatives clarified the public role in harnessing waterpower. He was sensitive to America's tradition of private enterprise. He stopped short of nationalizing electric power services. The federal government was given a mandate to transmit power from federal dams to non-federal entities for final distri-

---

[9] H.S. Person, "The Rural Electrification Administration in Perspective," *Agricultural History*, Vol. 24, No. 2, April 1950, p. 82.

[10] Hyman, *America's Electric Utilities*, p. 148.

bution to users. Preference was given to public entities, including municipalities, cooperatives, and public utility districts.

For years after Roosevelt became president, private utilities feared they would lose customers to federally supplied distribution systems. That uncertainty, coupled with federal subsidization of public utility alternatives, made it impossible to finance private waterpower facility construction during the 1930s. Wartime priorities further inhibited private facility construction through 1945. Big federal dams took center stage.

Roosevelt vanquished private power. He upended private utilities' domination of the industry. Between the time he became president in 1933 and the beginning of America's involvement in World War II in 1941, public hydroelectric capacity tripled to 2,750 megawatts. Meanwhile, private hydropower capacity remained essentially unchanged, rising only 3.2 percent, or 280 megawatts.[11] The federal government's share of total electricity production in the United States jumped from 5 percent in 1932 to almost 20 percent in 1944.

The scale, multiple use features, and social agenda of New Deal era federal projects were a vast departure from previous patterns in hydroelectric development. No private utility had the need or wherewithal to launch undertakings the size of these federal projects. Attention turned from electrifying America to harnessing America's waterways for irrigation, flood control, municipal water supply, and economic development—as well as for electricity. Politicians and federal agencies fought over dam building projects. Dam design and construction were institutionalized. The halcyon days of the 1920s were over.

---

[11] Federal Power Commission, *Electric Power Statistics, 1920–1940*, Washington, DC, 1941, pp. X–XII.

Fig. 15.1. Initial TVA Directors
Harcourt A. Morgan, Arthur E. Morgan, David E. Lilienthal (L to R)

Fig. 15.2. Wendell Willkie

Fig. 15.3. Lilienthal and Willkie Negotiations in March 1938

Fig. 15.4. TVA Rural Power Lines

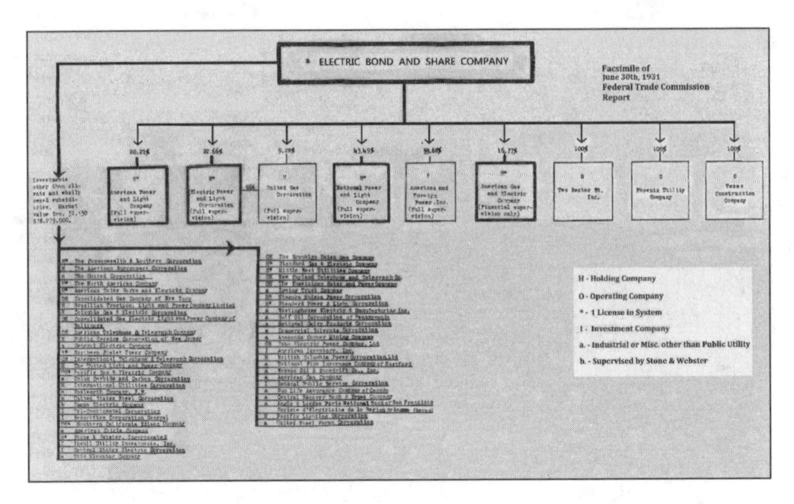

Fig. 15.5. Morgan-Controlled EBASCO, Largest Electric Holding Company Organization Chart in 1931

Fig. 15.6. Roosevelt Signs Public Utility Holding Company Act

Chapter Sixteen

That's the Story

Hydropower installations began in 1882, when a rudimentary direct-current system employing two 12.5 kW generators was installed in Appleton, Wisconsin, in order to light paper mills and illuminate the mansion of the mills' owner. This was only a month after Thomas Edison's first commercial system began lighting four hundred incandescent bulbs scattered around New York's financial district. The race was on to electrify America.

Electricity fundamentally altered life in America. It lit up the night. It powered factories and accelerated automation and product standardization. Electric streetcars displaced horses and carriages in cities nationwide. Meanwhile, the country's rapid population growth stoked demand. From 50 million at the time of Edison's first commercial system installation, the national population rose to 125 million fifty years later.

Savvy innovators, entrepreneurs, and promoters comprehended the magnitude of the opportunity to electrify America and seized it. The huge capital needs for success led to uneasy alliances with the bankers of the day. Mergers and acquisitions abounded as industry players jockeyed for scale and monopoly position. Politicians swarmed at every level.

From 1900–30, electrification drove the American economy. Electric utilities and their suppliers utilized more capital than any other industry and rivaled railroad investments of the late 1800s in terms of the relative share of gross national product. The large capital requirements and perceived stability of electric utility securities attracted the country's leading investment bankers.

Fifty years after Appleton and Edison's lower Manhattan installation, 70 percent of homes in the United States had electricity. Hydroelectricity was critical to that remarkable achievement. Where rivers and streams could be tapped, waterpower was the cheapest energy source. In the early 1900s, hydroelectric power accounted for more than 40 percent of the country's supply of electricity. Over fifteen hundred hydroelectric facilities produced about one third of the nation's electrical energy in 1940.

Because operations of direct-current systems were restricted to about a mile from their generators, early hydroelectric system installations evolved into a patchwork quilt of relatively small, dedicated, special-purpose systems and small-footprint community utilities located where abundant waterpower could be harnessed. System developers took advantage of existing hydromechanical power apparatus at mills and factories whenever possible.

The introduction of alternating-current systems led to explosive industry expansion from about 1900 on. With a relaxed footprint of operation and rapidly increasing demand, systems became larger and larger. Dams went from being 5–10 feet tall and hundreds of feet long to more than 700 feet high and over a mile long. Some dams created reservoirs that stretched for more than a hundred miles. The pressure for larger generator units continued to grow. In 1936, Hoover Dam's first 82.5 MW generator went into operation. Today, 800 MW generator units are available.

In addition to Edison's original discoveries and introduction of alternating-current generators, transformers, motors, and other AC system components, dam-related technical innovations were important to the evolution of hydropower. Among these were the development of analytic techniques upon which to base dam designs as well advances in concrete science. Learning how to control concrete's heat generation while it cured was especially critical to construction of massive, tall structures.

The dam-building knowledge base increased with each dam built. By the 1920s, ever-increasing standardization led to greater and greater efficiencies and risk reduction. Steam and gas-powered construction equipment eventually displaced mules and raw human labor.

Dams transformed their locale. Views of Hoover Dam's site before and after construction showed a striking contrast. Dams became an art form. Designers recognized the importance of visual impact. They wanted their dams to convey safety, permanence, and engineering prowess. Renowned architect Stanford White designed the first Niagara Falls powerhouse. The TVA's designs were modernistic sculptures.

As hydropower spread, many of the best dam sites were situated in remote, virtually inaccessible locations. Transportation of workers and materials to some construction sites was possible only by mule team. In 1910, for the mammoth Big Creek project high in California's Sierra Nevadas, a 56-mile-long construction railroad was built—with 1,100 sharp curves as it wound up the mountain sides to a connection with an incline railroad for the final 2,000-foot ascent.

The westward migration of people across the United States during the first half of the twentieth century stimulated the advent of large, multipurpose dams. The US Bureau of Reclamation played a major role in building dams in the West whose reservoirs could be used for irrigation, drinking water storage, and flood control while the projects were being paid for by the sale of the hydroelectricity they generated.

The nature of the symbiotic relationship between hydropower and America's rivers evolved over time. In the beginning, the primary concern was whether a dam would impede navigation. In 1908, President Theodore Roosevelt recognized the importance of simultaneously harnessing the country's rivers and expanding hydroelectric power:

> It is poor business to develop a river for navigation in such a way as to prevent its use for power, when by a little foresight it could be made to serve both purposes. We cannot afford needlessly to sacrifice power to irrigation, or irrigation to domestic water supply, when by taking thought we may have all three. Every stream should be used to the utmost.[1]

By the 1930s, hydropower was just one element of plans the federal government developed and implemented for comprehensive management of the water resources of entire river basins.

By the end of World War II, the golden era of hydropower that had captured the imagination of America ended. We had passed from electrifying the country to living in an electrified world—a world where electricity remains a readily available commodity and is taken for granted. At the same time, the nuclear age was dawning, and nuclear power soon would join waterpower and coal as a significant energy source. Even so, feder-

---

1 Message from President Roosevelt transmitting *A Preliminary Report of the Inland Waterways Commission* to Congress, February 26, 1908.

ally funded multipurpose dam projects continued to tame the nation's major rivers and turn them into series of lakes. These dams, in turn, created environmental and other concerns that over time increasingly throttled waterpower development.

As our story has unfolded, we encountered a fascinating montage of players. There was the prime mover Thomas Edison, as well as George Westinghouse and Nikola Tesla. The powerful banker J.P. Morgan, who always sought to build monopolies, was omnipresent. Among the industry consolidators were Henry Villard and Charles Coffin. Samuel Insull, now both underappreciated and defamed, was a major player in shaping and growing the electrical industry.

The remarkable pioneering entrepreneurs Charles Stone and Edwin Webster designed, built, owned, and/or managed a significant fraction of the country's hydroelectric capacity. Their epic confrontations in Washington State with Seattle's J.D. Ross were central to the war between private and public power.

Visionary Reclamation Director Arthur Davis conceived Hoover Dam and was the first to recommend construction of multipurpose dams whose powerplants would amortize total project costs. Hoover Dam made legendary dam builder Frank Crowe famous. It also launched the spectacular career of consummate entrepreneur Henry Kaiser.

The engineers crafting various hydroelectric projects often received little public attention. Notable exceptions were Philadelphia Electric's W.C.L. Eglin, Hugh Cooper of Keokuk and Wilson dam fame, and Reclamation's John Savage. Also lacking public attention were the hundreds of thousands of workers who actually built the dams and toiled namelessly. Most of them remain faceless in the pages of history. Yet the public has long marveled at their achievements.

A surprising force in the unfolding of the hydropower story was Henry Ford. His oft-unheralded impact was substantial. Inextricably intertwined with the story of Henry Ford was the story of Senator George Norris. Norris's crusade to bring public power to the Tennessee Valley was a tipping point in the public-versus-private power debate. The TVA's controversial first Chairman, Arthur Morgan, launched an unprecedented federal regional development initiative only to be fired by President Franklin D. Roosevelt as a result of a coup led by fellow director David Lilienthal.

Although the golden era of hydropower spanned the terms of thirteen US presidents, FDR far and away had the most impact. The tug-of-war between proponents of private power and public power that had begun in the early 1900s became increasingly brutal and bitter. Private power held the upper hand until the Great Depression—and Roosevelt's presidency. Roosevelt vanquished private power. He toppled the industry's kingpins. Large, multiple-purpose federal dam projects took center stage.

There is no doubt that dams will continue to contribute to humankind's efforts to control, harness, and manipulate the environment. Since hydropower is economical and provides about 7 percent of the total US electricity supply—and nearly half of all its renewable generation—it will remain an important element in the nation's energy supply equation. The focus going forward clearly will be upon upgrading the capacity of existing hydroelectric facilities and adding hydropower capability to existing dams that currently do not have it. Pumped storage will be a vitally important tool because it can enable grid flexibility and greater integration of variable generation resources such as solar and wind.

A comprehensive 2016 Department of Energy study concluded that combined US hydroelectric generating and storage capacity could be grown 50 percent by 2050, with 25 percent of the growth coming from new hydropower generation capacity (upgrades to existing plants, adding power to existing unpowered dams and canals, and limited development of presently undeveloped sites) and with 75 percent of the growth from new pumped storage capacity.[2] New hydroelectric dams will tend to be smaller and address specific, more localized needs. In this sense, hydropower has gone full cycle since its genesis in the 1880s.

Our story has focused on the golden era of dam building: waterpower's startup and rapid growth stages. Dams built then changed the map of the United States and vastly improved the lives of Americans. A century later, most of those dams still steadily produce clean, renewable energy.

Mural in US Department of the Interior Headquarters Building,Washington, DC.
Construction of a Dam by William Gropper, 1939

---

[2] US Department of Energy, *Hydropower Vision: A New Chapter for America's 1st Renewable Electricity Source*, Washington, DC, July 2016.

Afterword

George Jessup—A Dam Engineer

I have lived around dam people my entire life. My grandfather, George Penney Jessup, like Frank Crowe of Hoover Dam fame, was a builder. Between 1907 and 1952, Jessup managed the construction of twelve hydroelectric projects and four steam power plants. Working in the field to transform visions and plans into reality excited him. He had an uncanny ability to simultaneously comprehend myriad details and translate them into efficient work plans. Every project he managed was completed within budget and ahead of schedule.

Jessup descended directly from early eastern Long Island settlers. The initial Jessup family settler was a recorded landholder in Southampton as early as 1644. Jessup's father, Silas, was a leading citizen of Quogue, an oceanside resort village near Southampton where the family had had large land holdings since the 1600s. At age twenty-two, Silas became the youngest captain in the US Life Saving Service (now the Coast Guard) and commanded the Quogue Station. He also was an entrepreneur and a volunteer fireman.

On February 29 in the leap year of 1884, Jessup and his twin sister were born during a furious snowstorm in a house close to the Quogue Life Saving Station. This was less than one and a half years after the first hydroelectric system installation in Appleton and just months before Charlie Stone and Ted Webster became students at MIT. The Jessup twins had one older sister.

Jessup idolized his father and spent every moment he could with him, soaking up knowledge and working alongside him. In January 1900, while Jessup was a student at Blair Academy in New Jersey, he received a telegram that dramatically changed his life: Silas had died from pneumonia contracted after battling a fire in the pine forests near Quogue and then participating in a shipwreck rescue. Suddenly, Jessup, not quite sixteen years old, literally became the man of the family. It was a responsibility he took seriously for the rest of his life.

After graduating from high school, he became a student at Cornell University, where, in 1908, he earned a degree in Mechanical Engineering. He cofounded the Cornell chapter of the Acacia social fraternity. He served as a volunteer campus firefighter and was awarded a senior-year scholarship for his actions in combatting two major blazes. He worked part-time firing steam engines on the Lackawanna Railroad.[1] At the time of his first introduction to hydropower, in 1907, he was serving as a foreman on the construction of the University's hydroelectric power plant. This 1.9 MW facility generated power for the cam-

[1] While tending a train at a stop in Elmira, New York, Jessup tossed a bucketful of water overboard. To his dismay, the water drenched someone below. The person turned out to be Mark Twain. Jessup told me that the famous author laughed it off and a lengthy, fascinating conversation ensued.

pus, was used for teaching, and was a valuable research tool for testing and developing hydropower turbines.

After graduating from Cornell, Jessup managed construction projects for British and Canadian firms for several years before joining Stone & Webster, where his career turned to electricity. He participated in appraisals of electric street railways in New York City and in St. Augustine and other Florida cities. Thereafter, he focused exclusively on power facility construction. He supervised construction of steam power plants in Massachusetts, Virginia, Wyoming, and Puerto Rico. In these pages, we learned of his Stone & Webster hydroelectric projects at Iron Mountain, Michigan; Baker River, Washington; and Conowingo, Maryland.

Another major Stone & Webster project—Jessup's last while employed there—was Bagnell Dam on the Osage River in Missouri. Bagnell's reservoir formed the Lake of the Ozarks (see Figure 6.2). Bagnell project construction began in August 1929. Although the Great Depression had started two months later, work on the project continued. It became one of the country's only major ongoing construction projects during the hard times ahead, attracting thousands of workers from across the United States.

In 1932, with Bagnell completed and Stone & Webster reeling from the Depression, Jessup fell out of work despite by then being a nationally and an internationally known dam builder. He moved his family back to his ancestral home on Long Island and hunkered down to await an upturn in the economy. More than two difficult years later, the TVA reached out to him. He accepted the challenge to manage the completion of Wheeler Dam and then the construction of Guntersville and Kentucky dams. We saw that the end of World War II and the completion of

Kentucky Dam marked the end of the TVA's dam building era. Perhaps Jessup could have stayed with the TVA in a senior position, but an office job in a bureaucracy held no appeal for him. In 1946, Jessup left the TVA and became a highly sought-after hydroelectric consultant.[2]

After leaving the TVA, Jessup first supervised major repairs to Kentucky Utilities' Dix Dam on the Dix River near Harrodsburg, Kentucky. The 30 MW hydropower facility had been built in 1923–25 by Insull's Middle West Utilities. When constructed, the dam was the largest rock-filled dam in the world and the highest dam east of the Rockies. The dam, at 287 feet tall, was almost the same height as the 293-foot-high Baker River Dam built of concrete, which Jessup had completed in Washington State the same year. Jessup was engaged to tackle the persistent leakage issues that had plagued the dam since it initially was built.

On behalf of the Lower Colorado River Authority, he oversaw construction of two hydroelectric dams built in tandem on the Colorado River northwest of Austin to supply energy to central Texas. These were the Granite Shoals Dam (later renamed Wirtz Dam), which formed Lake LBJ, and the farther downstream Marble Falls Dam (renamed Starcke Dam). The dams, with combined electric capacity of 101 MW, were completed in 1951. There was an Insull connection here, too: An Insull

---

[2] In 1945, there supposedly was talk within the TVA that Jessup was being considered for a high-level position upon completion of the Kentucky project. Such talk apparently ceased after Jessup's encounter with Truman in October 1945. It is not clear that he would have accepted such a position if offered. The TVA underwent major changes in 1946. Chairman Lilienthal and TVA's other significant dam builder, Fred Schlemmer, also departed that year.

subsidiary had begun construction in 1931 of the 2-mile-long Hamilton Dam (now the Buchanan Dam) across the Colorado near Burnet, Texas. The project foundered when Insull's empire collapsed. The LCRA was established in 1934 to access federal funding from the Public Works Administration to complete the construction. Buchanan Dam became the first of six dams constructed by the LCRA to form the chain of reservoirs known as the Highland Lakes. Wirtz and Starcke were the final two completed.

Jessup's last project prior to retiring was for the construction of Consolidated Mining and Smelting Company of Canada's (now Teck Cominco) Waneta Dam to provide additional power for its giant lead and zinc smelter at Trail, British Columbia. The dam is located near the mouth of the Pend d'Oreille River as it joins the Columbia River less than a mile from the Canada–United States border. Construction began in 1951 and was completed at the end of 1953. The 249-foot-high dam is now owned by BC Hydro and has a rated capacity of 490 MW.

Jessup hated office work. His preferred office was at the project site. He was used to being in charge. He made things happen—and happen right. He was notoriously hands-on and seemingly omnipresent. For the Bagnell Dam project, his office was in an observation tower with windows and a deck all around, built on a rock outcropping high above the construction site. He could walk the deck and see everything below, day and night, through his high-powered field glasses. He put a cot inside and lived and worked there around the clock for long periods. If he saw something amiss, he would grab his telephone and forcefully reprimand an unsuspecting foreman or supervisor.

Similarly, for the Kentucky Dam project, his office was on a knoll above the job site. The walls facing the construction site had large windows, and there was a walkway just outside. Jessup would pace and survey the construction through his field glasses, his telephone nearby. He frequently would jump in his car and ride all through the project. If something was wrong, he would be right there talking to the foreman. As land was being cleared upriver for the reservoir, he would travel by motorboat to check on progress. On all his projects, he periodically would punch in incognito for a shift on a work crew to get a sense of what was happening and what the laborers were thinking. He knew what was going on.

My mother, Jessup's daughter, came home from her first day at school in first grade and reported to her father that her teacher had asked what her daddy did. She hadn't known what to say. Jessup harrumphed and said, "Tell her I am a dam engineer and proud of it!" It was true, but, in reality, he was a field engineer, a builder. Design engineers, using their analytical tools, defined each dam in detail. Jessup took what the design engineers had committed to paper and transformed it into concrete, turbines, and generators.

Jessup was practical. A personal experience makes the point. When I was living with him in high school, I took a physics class. My homework included a basic fulcrum (or teeter-totter) problem: If a person weighing X pounds sat some distance away from the balance point on one side, how far away from the balance point should someone weighing Y sit on the other side? He said, "I don't know the damn formula. Let's figure it out."

We went into his workshop, where he found a board and a box containing some identical parts. We made a seesaw with the board and placed a part on one side. We then placed a part on the other side. The board balanced when the distance of both parts from the center was the same. Then we put two parts on the second side and found the distance at which the board balanced again. After more iterations with more parts on the second side, we had the formula: Weight times distance equals weight times distance. I never have forgotten it!

Another example vividly demonstrates Jessup's belief that an academic foundation needs to be supplemented with first-hand knowledge and practical experience. Each year, the TVA hired a number of engineers graduating from top-rated colleges around the country. Those sent to Jessup's projects found themselves assigned to work crews. If they complained that they were engineers not laborers, Jessup delivered a succinct message: "When you can come back and tell me how that crew can do a better job, I will think of you as an engineer."

Jessup held himself to the same standard. He always was looking for ways to do things better. He religiously read trade magazines such as *Engineering News-Record,* both the articles and the advertisements (he thought it took way too long for new ideas to find their way into technical journal articles). He communicated widely within Stone & Webster and the TVA to learn of new innovations. He was active in the National Society of Professional Engineers and the American Society of Mechanical Engineers.

Not only was Jessup practical and hands-on, he also was a visual person. He paced while looking through his field glasses. He had precise clay models of his projects built on platforms about 10 feet long. The platforms were attached to chains whereby the model could be raised to the construction-office ceiling for storage and lowered to discuss some project particular. When he ventured forth by foot, automobile, or motor launch, he made a point of trying never to take exactly the same way twice.

Jessup had a reputation. Workers on his projects variously described him as aloof, domineering, demanding, gruff, impersonal, tough but fair. He was feared by many but respected by all. He had a strong work ethic and expected his workers to have one, too. He was a firm disciplinarian. If he saw a work crew shirking and thought they were not doing their work, he would lay into them, loudly and directly. His sentences were sprinkled liberally with "damn" and "hell" (the only curse words he used).

Make no mistake about it: Jessup cared about his workers. He strongly defended the workers on his projects. On one of his early projects, he pointed out to the superintendent that having a veterinarian on staff had greatly reduced the non-productive time of the work mules on the job. Why couldn't they employ a doctor on-site to look after the men? One was hired. During the Great Depression, men swarmed to the Bagnell project from all over the country desperate for jobs. Those not hired were sent to the cafeteria for a free meal. Jessup set an edict that nobody was to leave hungry.

Workers who did their job fared well and became loyal to Jessup. Many of the foremen and supervisors followed him from project to project. That loyalty lasted a long time. When I was young, a stranger knocked on our front door and asked for my mother. The man said that he had learned my mother was Jessup's daughter, and he wanted her to know that he thought Jessup was the finest man he ever had worked for. He had worked on three projects for Jessup. He explained that he had lost track of Jessup and asked her to pass along his respects.

It should be noted that many people who worked with him found over time that, underneath Jessup's gruff exterior, was a humorous, kind man. I myself discovered that when I lived with him—and I wasn't the only one. In tributes after his death, Paducah's newspaper summed up Jessup this way:

> He gained the name of a rough, tough, construction boss who demanded an all-out effort 24 hours a day. But he had another side—one of quick compassion and a keen sense of humor. Many of the thousands of men who worked under him knew both sides, and Jessup became highly popular with the mass of men who built [Kentucky Dam] .... He left a mark on Western Kentucky that will be long remembered. The dam itself is a monument to his ability as a construction engineer. He supervised its construction from start to finish.... Mr. Jessup would not claim major credit for the biggest of the TVA projects. There were many sub-bosses and others who had important parts in the work. But he welded these other men and the construction crews into the efficient team that put the vast project together without any major mishap. In the doing, he created the favorable image for TVA that developed as the Kentucky Dam work got underway.[3,4]

Even today, TVA historians continue to laud a humanitarian mission initiated by Jessup. As mentioned in Chapter Fourteen, severe Ohio River flooding in January 1937 submerged Paducah, Kentucky, and caused the worst natural disaster in the city's history. The TVA learned on January 22 that rising floodwaters had breached the earthen levee protecting Paducah and were inundating the city. Jessup, who then was managing the construction of Wheeler and Guntersville dams, quickly began organizing a TVA task force to render assistance to people marooned in Paducah.

With the TVA's steam towboat *Hiwassee* in the lead, Jessup,

---

3 "G. Jessup, Head Man at Dam, Dies," *Paducah Sun-Democrat,* October 1, 1961, p. 1.

4 "Jessup Well Remembered," *Paducah Sun-Democrat,* October 3, 1961, p. 4.

3 barges, 48 men, and 22 inboard and outboard mosquito boats normally used for malaria control traveled the 275 miles downstream from Wheeler to Paducah. The fleet arrived in the morning of January 25 after navigating by moonlight the previous night. The situation in Paducah was dire. About 95 percent of the city was inundated, and cold, debris-filled water extended up to 3 miles or so inland—and the water was continuing to rise.

Thousands of people had taken refuge in churches, schoolhouses, the railroad station, and other downtown commercial buildings. These locations, too, were being submerged and were crowded to a dangerous point. Panic was setting in. Jessup immediately put his twenty-two mosquito boats in to around-the-clock operation evacuating stranded residents.

The river crested on February 2 at about 11 feet above flood stage. Jessup's fleet continued to operate through February 6. Local sources estimated that around twenty thousand refugees were handled by the TVA boats.

At one point, a boat carrying Jessup even rescued a rooster stranded on a rooftop. The rooster hopped aboard the *Hiwassee* upon the boat's return and could not be dislodged. Eventually Jessup adopted him, named him Paducah Bill, and made him the Guntersville construction project's mascot. Jessup occasionally could be seen walking the site holding Paducah Bill. Today the rooster is memorialized on the TVA's website.[5]

The people of Paducah remained grateful for the TVA's rescue assistance during the 1937 flood and welcomed Jessup when he returned a year and a half later as project manager for the construction of Kentucky Dam.

---

[5] See, for example, www.tva.gov/About-TVA/Our-History/Built-for-the-People/Paducah-Bill-Moves-to-Guntersville.

The National Safety Council frequently recognized Jessup for his pioneering efforts to ensure his project sites were safe places to work and to reduce time lost by workers due to accidents and illness. During his days at the TVA, his projects were cited multiple times as having the best record in the country for accident frequency among dam construction projects. And, unlike Frank Crowe's Hoover Dam project, where more than two hundred men died, Jessup's TVA projects averaged fewer than ten fatalities. This was thanks in no small part to one of Jessup's innovations: offering first-aid classes to his workers. This training led workers to think about preventing accidents that would make first aid necessary. It also made critical assistance available more quickly when accidents did occur on a far-flung construction site. More than two thousand employees were trained on the Kentucky Dam project alone.

Given his active nature, Jessup was miserable when he retired from being a hydroelectric consultant after the Waneta Dam project in British Columbia. His wife and he had moved back to Ithaca, New York, to be near Cornell University again. He had been fiercely devoted to Cornell ever since his college days. All four of his children attended Cornell. After about a month of boredom, Jessup went to the personnel office at Cornell and asked for a job. He said he would do anything, even be a night watchman. The next day he received a call from the University's president, who told him that Cornell was embarking on a major campus expansion program and had nobody to oversee construction on the University's behalf. Among the projects were new residence halls and a new undergraduate library. Would Jessup be interested? Needless to say, he jumped at the opportunity. Jessup served as project engineer on university construction projects until he was felled by cancer in 1961. In an unusual move, the University posthumously named a busy campus thoroughfare Jessup Road in his honor.

My grandfather and I had a number of serious grandfather–grandson talks while I lived with him in high school. One of those, just months before he died, was especially memorable. He said that all of us have a duty during our life to try to make the world a little bit better for having been here: "You don't have to be Thomas Edison or Henry Ford ... even little things matter." Then he told me that his dams were his legacy, that he believed they helped make many people's lives better, and that he hoped this legacy would last long after he died. I believe he accomplished his goal.

Fig. AW.1. Jessup Upon Cornell Graduation in 1908

Fig. AW.2. Cornell University Hydropower Plant

Fig. AW.3. Bagnell Dam

Fig. AW.4. Jessup During Bagnell Dam Construction

Fig. AW.5. Jessup with Signature Hat and Pipe

Fig. AW.6. Dix Dam in 1928

Fig. AW.7. Wirtz Dam

Fig. AW.8. Starcke Dam

Fig. AW.9. Waneta Dam

Fig. AW.10. George Jessup Road Dedication Plaque

Appendix

Wheeler Dam Inspection Team

Wilfred M. Hall (1894–1986). Construction Engineer, Wheeler Dam, TVA. Among the construction projects he led prior to joining the TVA late in 1933 was Candlewood Dam in Connecticut. He went on to become the CEO of Charles T. Main, a major construction and engineering firm, and a member of the National Academy of Engineering. He earned his B.S. in Civil Engineering at the University of Colorado and was awarded an honorary Doctor of Engineering degree by Tufts University. By the time of his death, he was called the "Grand Old Man of Engineering."

Sherman M. Woodward (1871–1953). Chief Water Control Planning Engineer, TVA. Until joining the TVA in 1934, he had for many years been a professor of mechanics and hydraulics at Iowa State University, during which time he consulted widely on water control projects. He was especially known for his work with the Miami Conservancy District in shaping its Miami Valley flood protection program, where he worked closely with A.E. Morgan.

Llewellyn Evans (1882–1975). Chief Electrical Engineer, TVA. Prior to joining the TVA in 1933, he had been superintendent of Tacoma Power, which erected the Cushman dams and was at the forefront of the public power movement. At the TVA, he reported directly to Lilienthal, headed the Power Division, and was in charge of electricity market, systems and rates planning. He held a B.S. in Electrical Engineering from the University of California.

Leslie N. McClellan (1888–1981). Engineering consultant. Chief Electrical Engineer, US Bureau of Reclamation. He later became Reclamation's Chief Engineer and was a key contributor to the Hoover Dam and Grand Coulee projects. He earned his B.S. in Electrical Engineering from the University of Southern California and was awarded an honorary Doctor of Engineering degree by the University of Colorado.

Charles H. Paul (1875–1941). Engineering consultant. With Reclamation early in his career, he was in charge of the construction of Arrowrock Dam. He later served as chief engineer of the Miami Conservancy District. A widely known civil engineer, he became an independent consultant in 1925, remaining a consulting engineer to the District, the TVA, and Reclamation until his sudden death in 1941. He was engaged extensively in Reclamation's Grand Coulee project. He was educated at MIT.

John L. Savage (1879–1967). Engineering consultant. Chief Designing Engineer, US Bureau of Reclamation. A world-renowned expert on dams and civil engineering, he was in charge of all civil, electrical, and mechanical design for Reclamation. His projects included Hoover, Shasta, and Grand Coulee dams. So that construction of Norris and Wheeler dams could begin as quickly as possible after the formation of TVA, Reclamation, under Savage's leadership, was engaged to prepare detailed designs and contract drawings for those dams. He received his B.S. in Civil Engineering from the University of Wisconsin and was awarded honorary doctorates by the University of Wisconsin, the University of Denver, and the University of Colorado. He was a member of the National Academy of Sciences.

Charles H. Locher (1862–1948). Engineering consultant. He was a highly experienced and diversified construction contractor. He managed construction of, for example, Shoshone Dam, portions of the Catskill Aqueduct, and the entire Miami Valley flood protection project. He consulted on construction methods for Norris, Wheeler, and Pickwick dams. Although he had no formal education after high school, he was awarded an honorary Doctor of Engineering degree by the University of Maryland.

George P. Jessup (1884–1961). Construction Superintendent, Wheeler Dam, TVA. He previously had managed the construction of Baker Dam (Concrete, Washington), the Conowingo hydroelectric facility on the Susquehanna River, and Bagnell Dam (forming the Lake of the Ozarks). He subsequently managed the construction of the TVA's Guntersville and Kentucky dams. He received his B.S. in Mechanical Engineering from Cornell University.

Carl A. Bock (1883–1966). Assistant Chief Engineer, TVA. He was a civil engineer with degrees from Cornell College in Iowa and was the first engineer hired by the TVA. A.E. Morgan had relied heavily on Bock when Bock worked for him during the Miami Valley flood protection program, and he continued to do so after the formation of the TVA. After Morgan was fired by Franklin D. Roosevelt, Bock was shunted aside and resigned from the TVA in 1939. He later became chief engineer and executive director of the Puerto Rico Water Resources Authority.

Leroy F. Harza (1882–1953). Engineering consultant. An internationally famous consultant on dams and hydroelectric power plants, he received civil engineering degrees from the University of Wisconsin. He founded Harza Engineering Company in Chicago in 1920 with the assistance of Samuel Insull. Over the years, he designed more than forty hydroelectric projects and was consulted by such agencies as the TVA and the Corps of Engineers on many more. Among his more important projects were Dix Dam in Kentucky, the Santee-Cooper Project in South Carolina, the International Rapids Hydro Project on the St. Lawrence River, and the Fort Peck power installations on the Missouri River. He was largely responsible for the pioneering installation of Kaplan turbines at Bonneville Dam.

Lt. Col. Charles E. Perry (1879–1938). Nashville District Engineer, US Army Corps of Engineers. Commanding officer of the Corps district that had responsibility for navigation on the Tennessee River. The Corps designed and constructed the lock for Wheeler Dam (the Corps designed the locks for all TVA dams on the Tennessee River except Kentucky Dam). He received his civil engineering education at Yale University. Tragically, Perry drowned in March 1938 after falling from the spillway wall at Wheeler Dam during an official inspection visit. He was buried in Arlington National Cemetery.

Acknowledgments

One evening I was regaling my son-in-law, Brian Murphy, with stories of my grandfather and his dams. Brian said, "You should write a book." The thought never had crossed my mind, and I replied, "No way." Several months later, he again urged me to do so. I eventually decided it would be an interesting and worthwhile project.

I quickly learned that writing a book is a fascinating journey involving thousands of hours of research. I could not have tackled a book of this scope twenty years ago. Since then, Google and other parties have digitally preserved and made available online countless old, sometimes obscure, books, newspapers, journal articles, photographs, and other reference materials. This has been a true gift. How else would I have been able to find the menu from a celebratory luncheon marking the completion of Baker Dam in 1925?

Book writing exposed my proclivity for long sentences and run-on paragraphs. My wonderful editor, Kelli Christiansen, reined me in and made countless suggestions that markedly improved my initial draft.

The book contains 165 photographs, many rare. My permissions editor, Karen Ehrmann, is a miracle worker. She confirmed clearance to use every photo. What an accomplishment!

Publishing a book was new to me. Jim Kepler of Adams Press patiently encouraged and tutored me. Without him, there would be no book.

Speaking of patience, my wife, Susan, wins a gold star. Her understanding carried me through the many roadblocks and dead ends that cropped up along the way. I am deeply grateful for her love and support.

My family and friends have been intrigued by my efforts. Their continuing enthusiasm has amazed me, buoyed me, and propelled me to finish the manuscript and get the book, my son-in-law's brainchild, published.

I always have greatly admired and respected my grandfather. I sensed his presence constantly as I worked on this book. It is my tribute to him and my gift to his memory. I believe he would approve.

Illustration Credits

FRONT COVER. Photo by Gabriel Moulin. Library of Congress, Item 2005677820.

FIG. 1.1. Yale Collection of Western Americana, Beinecke Rare Book and Manuscript Library, Yale University.

FIG. 1.2. Original source unknown.

FIG. 1.3. Photo by Lewis Wickes Hine. National Archives and Records Administration, NAID 523307.

FIG. 1.4. Photo by Solomon D. Butcher (Solomon Devore). Library of Congress Prints and Photographs Division, LOC 2005693386.

FIG. 1.5. Photo by J.S. Johnston. Library of Congress Prints and Photographs Division, LCCN 2004682037.

FIG. 1.6. Photo by Lewis Wickes Hine. National Child Labor Committee Collection, Library of Congress Prints and Photographs Division, LOC 2018677059.

FIG. 2.1. GL Archive/ Alamy Stock Photo.

FIG. 2.2. miSci (Museum of Innovation and Science).

FIG. 2.3. *Electricity*, Vol. 4, No. 15, April 26, 1893, p. 202.

FIG. 2.4. Consolidated Edison Company of New York. Artist unknown.

FIG. 2.5. Hearthstone Historic House Museum.

FIG. 2.6. Used by permission of WEC Energy Group.

FIG. 3.1. Created by author.

FIG. 3.2. Created by author.

FIG. 3.3. Photograph by Joseph G. Gessford (1902). George Westinghouse Museum Collection, Detre Library & Archives, Heinz History Center, Identifier 20170323-hpichswp-0006.

FIG. 3.4. Henry Villard, *Memoirs of Henry Villard, Journalist and Financier, 1835–1900, Volume I*, Cambridge, MA: The Riverside Press, 1904, frontispiece photograph by W. Höffert. Courtesy of the Internet Archive.

FIG. 3.5. Photographer unknown.

FIG. 3.6. Print collection, The Miriam and Ira D. Wallach Division of Art, Prints and Photographs, The New York Public Library, 1701402.

FIG. 3.7. FLHC 3/ Alamy Stock Photo.

FIG. 3.8. Photographer unknown.

FIG. 3.9. Richard Gunderman, "Nikola Tesla: The extraordinary life of a modern Prometheus," *The Conversation*, January 3, 2018. Photographer unknown.

FIG. 3.10. General photograph collection, Power Projects, The Buffalo History Museum.

FIG. 4.1. Granger Historical Picture Archive.

FIG. 4.2. Covers of November 29, 1926, and November 4, 1929, from *TIME Magazine* Archive. Cover of May 14, 1934, from *TIME* © May 14, 1934 TIME USA LLC, all rights reserved, used under license.

FIG. 4.3. Commonwealth Edison Company.

FIG. 4.4. *Electrical Merchandise and Selling Electricity*, Vol. 11, No. 8, August 1912, p. 341.

FIG. 4.5. Photographer unknown.

FIG. 4.6. Photo courtesy of Lyric Opera of Chicago.

FIG. 5.1. Photos courtesy of the United States Society on Dams and the Tennessee Valley Authority.

FIG. 5.2. Created by author.

FIG. 5.3. Xing Luo *et al.*, "Overview of Current Development in Electrical Energy Storage Technologies and the Application Potential in Power System Operation," *Applied Energy*, Vol. 137, January 2015, p. 514.

FIG. 5.4. Data from Energy Policy Institute of Chicago.

FIG. 5.5. US Geological Survey.

FIG. 5.6. California State Library Photo: Hervey Friend, 1891.

FIG. 5.7. Library of Congress, LC-DIG-ppmsca-17716.

FIG. 5.8. David Darling, *Encyclopedia of Alternative Energy and Sustainable Living*.

FIG. 5.9. US Bureau of Reclamation photo archives.

FIG. 6.1. Massachusetts Institute of Technology.

FIG. 6.2. Created by author using data compiled from multiple sources.

FIG. 6.3. US Census Bureau.

FIG. 6.4. Courtesy of MIT Museum.

FIG. 6.5. Courtesy of MIT Museum.

FIG. 6.6. Stone & Webster Engineering Corporation *Water Powers* brochure, 1913, plate 37. Courtesy of HathiTrust.

FIG. 6.7. *Ibid.*, p. 67.

FIG. 6.8. *Ibid.*, plate 33.

FIG. 6.9. *Ibid.*, plate 8.

FIG. 6.10. University of Washington Libraries, Special Collections, WAS0880.

FIG. 6.11. Public Service Company of Colorado. Original photographer unknown.

FIG. 7.1. Mississippi River Power Company.

FIG. 7.2. Used by permission of Ameren Missouri.

FIG. 7.3. Mississippi River Power Company. Photograph by H.M. Anschutz.

FIG. 7.4. George Grantham Bain Collection, Library of Congress, LC-DIG-ggbain-50130.

FIG. 7.5. *Engineering Record,* Vol. 64, No. 6, August 5, 1911, p. 150. Courtesy of HathiTrust.

FIG. 7.6. Mississippi River Power Company, Bulletin 8, December 1912, p. 14. Regional History Collection, Western Illinois University Libraries Archives and Special Collections.

FIG. 7.7. Mississippi River Power Company, Bulletin 10, July 1913, p. 4. Regional History Collection, Western Illinois University Libraries Archives and Special Collections.

FIG. 7.8. *Stone & Webster Journal,* Vol. 27, December 1920, p. 448. Courtesy of HathiTrust.

FIG. 7.9. *Ibid.,* p. 450.

FIG. 7.10. Mississippi River Power Company, Bulletin 10, July 1913, p. 2. Regional History Collection, Western Illinois University Libraries Archives and Special Collections.

FIG. 8.1. US Department of Energy.

FIG. 8.2. US Bureau of Reclamation.

FIG. 8.3. Library of Congress, HAER UTAH, 25-PAYS, 1–25.

FIG. 8.4. US Army Corps of Engineers, Detroit District.

FIG. 8.5. Paramount and Pathé Newsreel Collection, Sherman Grinberg Library.

FIG. 8.6. Tennessee Valley Authority.

FIG. 8.7. National Photo Company Collection, Library of Congress, LC-F81-11324.

FIG. 9.1. Created by author.

FIG. 9.2. Detroit Publishing Company, Library of Congress, LC 2016816959.

FIG. 9.3. Brian Callahan/ Flickr.

FIG. 9.4. Minnesota Historical Society.

FIG. 9.5. US Army Corps of Engineers, VIRIN: 160805-D-CR197-001.

FIG. 9.6. David Keller, *Stone & Webster 1889–1989: A Century of Integrity and Service*, New York: Stone & Webster, 1989, p. 121.

FIG. 9.7. Image from the Collections of The Henry Ford, Acc 1660, Box 153, Image #THF37614.

FIG. 9.8. Ford Motor Company.

FIG. 9.9. Photo courtesy of UpNorth Memories Postcards.

FIG. 9.10. Superior View Photography.

FIG. 9.11. Image from the Collections of The Henry Ford, Acc 1660, Box 64, Image #THF99978.

FIG. 9.12. Ford yacht photo from the Collections of The Henry Ford, Acc 1660, Box 112, Image # THF140396. Automobile caravan a video still from *Vagabonds*, footage of Ford and friends camping (1923), The Henry Ford.

FIG. 9.13. Image from the Collections of The Henry Ford, Acc 1660, Box 66, Image # THF101715.

FIG. 9.14. Image from the Collections of The Henry Ford, Digital Photo, 2004, Image # THF148103.

FIG. 10.1. Created by author from multiple sources.

FIG. 10.2. US Census Bureau.

FIG. 10.3. Puget Sound Power & Light Company 1924 Annual Report, University of Pennsylvania Library.

FIG. 10.4. Author's collection.

FIG. 10.5. Seattle Municipal Archives.

FIG. 10.6. University of Washington Libraries, Special Collections, WWDL0650.

FIG. 10.7. University of Washington Libraries, Special Collections, KHL047.

FIG. 10.8. Photo courtesy of Tacoma Power.

FIG. 10.9. Washington State Historical Society, 1998.62.9.

FIG. 10.10. Author's collection.

FIG. 10.11. Washington State Historical Society, 2002.59.56.

FIG. 10.12. Puget Sound Power & Light Company.
FIG. 10.13. Puget Sound Power & Light Company 1924 Annual Report, University of Pennsylvania Library.

FIG. 11.1. National Museum of American History, Smithsonian Institution, Image 80-16516.
FIG. 11.2. Nicholas Wainwright, *History of the Philadelphia Electric Company, 1881–1961*, Philadelphia, PA: Philadelphia Electric Company, 1961, after page 194. Courtesy of HathiTrust.
FIG. 11.3. Richmond Station in the 1920s, company photo no. 36774, PECO Library.
FIG. 11.4. Victor Dallin Aerial Survey Collection (1970.200), Audiovisual Collections and Digital Initiatives Department, Hagley Museum and Library, Wilmington, Delaware. Courtesy of the Hagley Museum and Library.
FIG. 11.5. Stone & Webster, *Conowingo*, 1928, p. 4. Photographer unknown.
FIG. 11.6. Theodor Horydczak Collection, Library of Congress Prints and Photographs Division, LC-H824-1184-006-B.
FIG. 11.7. Stone & Webster, *Conowingo*, p. 25. Photographer unknown.
FIG. 11.8. *Ibid.*, p. 29. Photographer unknown.
FIG. 11.9. Wainwright, *Philadelphia Electric*, after page 178. Courtesy of HathiTrust.
FIG. 11.10. Original photographer unknown (circa 1927).

FIG. 12.1. US Bureau of Reclamation.
FIG. 12.2. Photograph album collection, Library Special Collections, Charles E. Young Research Library, UCLA.
FIG. 12.3. Bain Collection, Library of Congress Prints and Photographs Division, LC-B2-118-12.
FIG. 12.4. US Bureau of Reclamation.
FIG. 12.5. *Los Angeles Times* staff. Copyright © 1935 *Los Angeles Times*. Used with permission.
FIG. 12.6. US Bureau of Reclamation.

FIG. 12.7. US Bureau of Reclamation.

FIG. 12.8. Photo by Ben D. Glaha for the US Bureau of Reclamation. Library of Congress Prints and Photographs Division, LC-DIG-ds-00633.

FIG. 12.9. Photo by Ben D. Glaha for the US Bureau of Reclamation. Library of Congress Prints and Photographs Division, LC-USZ62-114354.

FIG. 12.10. Keystone Photo Service/ Rare Historical Photos.

FIG. 12.11. US Bureau of Reclamation.

FIG. 13.1. William F. Willingham, *Army Engineers and the Development of Oregon: A History of the Portland District US Army Corps of Engineers,* US Army Corps of Engineers, 1983, p. 96.

FIG. 13.2. "Power, Navigation and Irrigation in Two Projects on the Columbia," *Engineering News-Record,* Vol. 113, No. 22, November 29, 1934, p. 678. Courtesy of the Internet Archive.

FIG. 13.3. Willingham, *History of the Portland District,* p. 104.

FIG. 13.4. US Farm Security Administration/Office of War Information. Library of Congress Prints and Photographs Division, LC-USF346- 070655-D.

FIG. 13.5. UPI/Newscom.

FIG. 13.6. US Bureau of Reclamation.

FIG. 13.7. US Bureau of Reclamation.

FIG. 13.8. US Bureau of Reclamation.

FIG. 14.1. Adapted from Tennessee Valley Authority.

FIG. 14.2. Photograph by G.L. Bracey for the Tennessee Valley Authority. Harold Clute Photograph Collection, Alabama Department of Archives and History, File Q42741.

FIG. 14.3. Data from various Tennessee Valley Authority publications.

FIG. 14.4. Tennessee Valley Authority.

FIG. 14.5. Photograph by Lewis Wickes Hine for the Tennessee Valley Authority. National Archives and Records Administration, NAID 532650.

FIG. 14.6. Tennessee Valley Authority.

FIG. 14.7. Tennessee Valley Authority.

FIG. 14.8. Tennessee Valley Authority.

FIG. 14.9. Records of the War Production Board. National Archives and Records Administration, NAID 534835 and NAID 515881.

FIG. 14.10. Tennessee Valley Authority.

FIG. 14.11. Records of the Tennessee Valley Authority, Record Group 142. National Archives at Atlanta, NAID 890185.

FIG. 14.12. Tennessee Valley Authority.

FIG. 14.13. Tennessee Valley Authority. Tennessee State Library and Archives, Image 23729.

FIG. 14.14. Photograph by Roland Wank for the Tennessee Valley Authority.

FIG. 14.15. Tennessee Valley Authority.

FIG. 14.16. Universal Newsreel, Volume 18, Release 441, October 11, 1945. MCA/Universal Pictures Collection 1929–67, National Archives and Records Administration, NAID 100520.

FIG. 14.17. Tennessee Valley Authority.

FIG. 15.1. Photograph by Granville Hunt for the Tennessee Valley Authority.

FIG. 15.2. Photo by Greystone Studio. Library of Congress Prints and Photographs Division, LC-USZ62-38331.

FIG. 15.3. Harris & Ewing Photograph Collection, Library of Congress Prints and Photographs Division, LC-H22-D-3487.

FIG. 15.4. Franklin D. Roosevelt Presidential Library and Museum, Photo ID 52333.

FIG. 15.5. US Federal Trade Commission.

FIG. 15.6. Everett Collection Historical/ Alamy Stock Photo.

CHAPTER 16 END PHOTO. US Department of the Interior.

FIG. AW.1. Author's collection.

FIG. AW.2. Frank Vetere/ Alamy Stock Photo.

FIG. AW.3. Courtesy of the Missouri State Archives.

FIG. AW.4. Author's collection.

FIG. AW.5. Author's collection.

FIG. AW.6. Caufield & Shook Collection, Photographic Archives, University of Louisville, ULPA CS 095346.

FIG. AW.7. Lower Colorado River Authority Corporate Archives, Image W00880.

FIG. AW.8. Lower Colorado River Authority Corporate Archives, Image W00481.

FIG. AW.9. Courtesy of the Trail Historical Society.

FIG. AW.10. Author's collection.

ABOUT THE AUTHOR. Personal photo from the author's collection.

Index

Page references in italics indicate figures.

Index

Bob Underwood was born in Paducah, Kentucky, as his grandfather was completing the construction of nearby Kentucky Dam. While Underwood was a child, his parents took him to see other dams whose construction his grandfather had supervised. Family dam-building stories captured his imagination. Later, as a Stanford engineer and Silicon Valley veteran with a career focused on developing technology-oriented businesses, he realized that the evolution of hydroelectric power rivaled that of any transformational technology seen in recent times. It was a story worth telling.

Underwood and his wife live in the Chicago area and summer on eastern Long Island in the same house where his dam-builder grandfather grew up.